Springer Series in Information Sciences   24

Editor: Teuvo Kohonen

Springer Series in Information Sciences

Editors: Thomas S. Huang   Teuvo Kohonen   Manfred R. Schroeder

Managing Editor: H.K.V. Lotsch

D.W. Heermann   A.N. Burkitt

Parallel Algorithms in Computational Science

With 46 Figures

Springer-Verlag

Berlin Heidelberg New York
London Paris Tokyo
Hong Kong Barcelona

Professor Dr. Dieter W. Heermann

Institut für Theoretische Physik der Universität Heidelberg, Philosophenweg 19,
D-6900 Heidelberg, Fed. Rep. of Germany

Anthony N. Burkitt, Ph.D.

Theoretische Physik: FB8, Gauss-Strasse 20, Bergische Universität – GHS,
D-5600 Wuppertal 1, Fed. Rep. of Germany

Series Editors:
Professor Thomas S. Huang

Department of Electrical Engineering and Coordinated Science Laboratory,
University of Illinois, Urbana, IL 61801, USA

Professor Teuvo Kohonen

Laboratory of Computer and Information Sciences, Helsinki University of Technology,
SF-02150 Espoo 15, Finland

Professor Dr. Manfred R. Schroeder

Drittes Physikalisches Institut, Universität Göttingen, Bürgerstrasse 42–44,
D-3400 Göttingen, Fed. Rep. of Germany

Managing Editor: Helmut K. V. Lotsch

Springer-Verlag, Tiergartenstrasse 17,
D-6900 Heidelberg, Fed. Rep. of Germany

ISBN-13:978-3-642-76267-3   e-ISBN-13:978-3-642-76265-9
DOI: 10.1007/978-3-642-76265-9

One split into many,
Many join to form one

One Thousand Airplanes on the Roof
Philipp Glass

Preface

Our aim in this book is to present and enlarge upon those aspects of parallel computing that are needed by practitioners of computational science. Today almost all classical sciences, such as mathematics, physics, chemistry and biology, employ numerical methods to help gain insight into nature. In addition to the traditional numerical methods, such as matrix inversions and the like, a whole new field of computational techniques has come to assume central importance, namely the numerical simulation methods. These methods are much less fully developed than those which are usually taught in a standard numerical mathematics course. However, they form a whole new set of tools for research in the physical sciences and are applicable to a very wide range of problems. At the same time there have been not only enormous strides forward in the speed and capability of computers but also dramatic new developments in computer architecture, and particularly in parallel computers. These improvements offer exciting prospects for computer studies of physical systems, and it is the new techniques and methods connected with such computer simulations that we seek to present in this book, particularly in the light of the possibilities opened up by parallel computers.

It is clearly not possible at this early stage to write a definitive book on simulation methods and parallel computing. Our contribution here is intended as just one of the first steps in the direction of producing parallel algorithms for simulation methods. The methods that we present are being rapidly refined and new methods are constantly being discovered.

The conventional boundaries between the sciences are today becoming blurred and losing their significance. Many concepts, methods and algorithms find their way from one discipline into another. The Monte Carlo and molecular dynamics methods, which are now widely used by physicists, chemists and biologists alike, are examples of such cross-fertilization. The techniques discussed in this book are thus of relevance beyond the field of computational physics to all who employ computer simulation methods to gain insight into the processes of nature.

We have tried to make our presentation comprehensible to a wide audience, but it is hardly possible to disguise our background in physics, from which we draw many examples. The techniques themselves remain relevant to other sciences, and the examples can be considered merely as stepping stones along the way.

This book results from a one-semester course which was given at the University of Wuppertal and from research which was undertaken by the authors. Thus the book is not primarily intended as a textbook for undergraduate or graduate students, but as a hybrid between a concise report on current research and a pedagogical introduction to parallel computing from the computational science point of view.

The reader is expected to be familiar with the basics of computer simulation methods. However, both for the reader's convenience and in order to introduce notation, the text contains a short introduction to the Monte Carlo and molecular dynamics methods. This does not and cannot replace a textbook on this subject. The reader will find ample references to recent up-to-date textbooks and articles in the corresponding sections.

In the first chapters we develop the ideas behind parallel algorithms by starting from problems in computational science and investigating their intrinsic parallelism. We then introduce the theoretical concepts necessary for their further development and investigation. This is followed by a chapter dealing primarily with the underlying hardware, in which we present the hardware concepts for parallel computation and give a number of examples of specialized hardware that has been built for specific problems in computational physics. We also discuss some general issues concerning languages for parallel computers.

Following these introductory chapters we start the discussion of specific parallel algorithms and parallelization methods for computational science problems. We present the general ideas behind the parallelization methods and then apply them to simulation problems, with particular emphasis on lattice and polymer systems. Many of the algorithms are presented in an Occam-like notation to facilitate both the readability and the convertibility into real programs. This notation is straightforward and needs little explanation, and any difficulties can be resolved by a quick reference to Chap. 9, in which Occam is discussed more fully.

Throughout the book the reader will observe that geometrically inspired parallelizations predominate. The reasons for this are manifold. Geometry plays an important role in our everyday experience and we have learnt almost intuitively how to relate objects to each other geometrically. More particularly, we have learnt when objects in space are independent of each other or influence one another only indirectly. It is thus natural that many of our basic ideas about parallelism have their origin in this experience and knowledge. Similarly, physics and the laws we use to describe physical phenomena are based on geometry. This geometric reduction, or partitioning, of systems is a key to carrying out calculations and performing simulations on parallel computers.

Non-geometrical methods are generally less straightforward. Much more research is needed into the nature of simulation methods and the physical problems themselves before parallelism can be exploited effectively on this level, although we present some simple methods that are useful.

The final chapter of this book includes a self-contained introduction to Occam. Although this language certainly has its limitations, we feel that it serves very well to introduce the concepts necessary for the understanding of parallelism in physics problems and in computer languages. For this reason the introduction to Occam is very pedagogic in character, a feature to which the language lends itself nicely. Secondly, the native language of transputers is Occam and the ready availability of transputers makes it possible to apply the concepts presented in this book to model and real-life problems.

It is a pleasure for us to thank those without whom this work would not have taken the shape that it has. Special thanks are due to D. Stauffer, K. Binder, K. Schilling, B. Forrest, W. Paul and P. Nielaba for critical reading of the manuscript and for numerous discussions.

We would like to thank the University of Wuppertal, and especially the members of the theory group of the Fachbereich Physik, for the warm working environment, which was a stimulus for us. One of us (DWH) is particularly grateful for the hospitality extended during the year 1988/89. Thanks are also due to the members of the condensed matter theory group at the University of Mainz, colleagues in the department of physics at Edinburgh University and many others who have provided stimulus and encouragement. We also thank the Gesellschaft für Mathematik und Datenverarbeitung at Schloß Birlinghoven and the Höchstleistungsrechenzentrum (HLRZ) for their support and hospitality.

We gratefully acknowledge financial support from the Sonderforschungsbereich 123 and 262, Schwerpunkt Schi/29, as well as the Bundesministerium für Forschung und Technologie, also BAYER under the grant 03M40284 and the German-Norwegian Cooperation Foundation under grant 26657D.

Heidelberg,   *Dieter W. Heermann*
Wuppertal, March 1990   *Anthony N. Burkitt*

Contents

1. Introduction

It has become almost a tautology that the computational requirements of science and engineering demand computational resources orders of magnitude larger than currently available with serial or vector machines! Even although it will be possible to increase the power of single processor machines significantly, perhaps by one order of magnitude or more, this is still not enough to make the solution of some science and engineering problems feasible with such machines. The necessary accuracy or the complexity of a problem frequently demands resources beyond the capacity of present and forseeable scalar or vector machines.

An alternative to seeking ever increasing power for a single processor is to use many processors at the same time to solve a problem. Such parallel machines exist today and are being used to study a wide variety of science problems. Some of these machines do not at present surpass the performance figures of the best conventional single-processor machines, but there are many problems which are well suited to parallel computation and the performance of a current-generation parallel machine with that of a suitable parallel algorithm can be comparable with that of a serial or vector supercomputer.

What is our aim with parallel computing? By applying more and more processors to the solution of a problem we can expect to decrease the time it takes to reach the solution. We certainly hope that the time taken does not increase, as typically happens with bureaucracy where the overhead (i.e., the time spent doing unproductive work) is large. Most useful types of parallel computing require the exchange of information between the processors. Assuming that we can reduce the adminstrative overhead sufficiently, the ideal that we can aim for is that the time a particular problem takes decreases linearly with the number of applied processors. In other words the speed-up of the time taken using a number of processors compared with a single processor machine should be linear. This ideal situation is shown in Fig. 1.1.

Let us investigate this idea a bit further. If we apply ten processors to solve a problem then we will finish ten times faster than using just one, if the speed-up is linear. Suppose we are bold and apply ten thousand processors. Can we expect still to find a linear speed-up and obtain the result ten thousand times faster? Is it better to use many processors with a limited capability or should we increase the power of each processor by a factor of ten and reduce the number of processors?

To address such questions we need to investigate parallelism, and particularly the amount of parallelism that is inherent in the system which we want to study. We also need to investigate the parallelism which the method or algorithm itself allows.

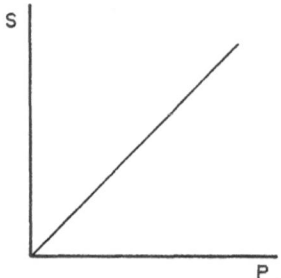

Fig. 1.1. The linear curve represents the best acceleration obtainable with an algorithm on a parallel computer. The number of processors is shown on the horizontal axis and the speed-up vertically

But what does the study of parallelism bring to science apart from the computational power and acceleration of algorithms? Parallelism brings with it a fresh view into science and physical processes. It opens new perspectives on physics and other sciences as well as the methods which we use to describe and formulate nature. Studying the inherent parallelism in the models and methods we sometimes discover another facet not apparent when looking at the problem from the usual angle.

Let us elaborate on what we understand as computational science. It is perhaps best to say right from the start that we understand computational physics as simulational physics, and by the same token computational science means simulational science. We want to study nature by making a model and following the development of the phenomena which the model exhibits in both space and time. Simulational science is thus more than numerical science.

In simulational science we use some method to generate states, which in general we denote by x, from a state space (or *phase space*) Ω. At the outset we assume that there are some particular given entities and a description of their interaction. The entities may be cellular automata, atoms, polymers and so forth. The state space Ω is the set of states x which such a system can assume.

The generated states should be representative of the parameters $p_1, ..., p_m$ under which we want to investigate the system. The parameters can, for example, correspond to different temperatures for a scan of the phase diagram, various system sizes or different concentrations of atoms. During a simulation a set of states

$$x_1, x_2, ..., x_n, \quad x_i \in \Omega \quad , \tag{1.1}$$

is generated with the general algorithm $\mathcal{A}$

```
x = x₀ ∈ Ω
FOR n
    change state xᵢ → xᵢ₊₁ ∈ Ω
ENDFOR
```

where x_0 is some appropriate initial state.

The problem which we wish to address is as follows. We want to generate a new state x_{i+1} from the state x_i in parallel

$$x_i \xrightarrow{\text{in parallel}} x_{i+1} \quad . \tag{1.2}$$

The statement *in parallel* means that we look for a reduction in the complexity of the generation of the new state over that on a serial machine. The reduction in the computational complexity of the function which leads from one state to the next will come through the use of many processors which can operate concurrently on the generation of a new state.

The arrow in (1.2) denotes a generic method. The methods of generating one state from another which we consider primarily are the Monte Carlo and molecular dynamics methods. These methods are used for illustration, although most of our results and investigations are more generally applicable.

We must, of course, limit our scope somewhat. In principle there exists the possibility of also considering the outer loop as a possible source of parallelism, which could bring a further reduction in the computational complexity. We could generate the set $\{x_1, ..., x_n\}$ *in parallel* instead of just looking at the generation of one state from another. For completeness we note the possibility of the parallel generation of the entire set

$$\text{generate in parallel } \{x_1, ..., x_n | x_i \in \Omega\} \quad . \tag{1.3}$$

Before we jump into further investigations of parallelism, let us briefly pause and think about possible limitations. That parallel computation is not the solution to all of our computational problems can be most impressively demonstrated by an almost trivial problem which does not involve any simulation. Consider raising a to the kth power. Any reasonable parallel algorithm takes exactly the same amount of time as it would on a serial computer. The complexity is the same for both types of computers.

The possibility that there does not exist a better solution using a parallel computer raises several questions. Having demonstrated that problems exist where the complexity cannot be reduced, can we identify more such problems? Can we classify them in any way? From a more general viewpoint we can go even further and ask if there are problems which are hard to parallelize. Later on we will be more specific about what we mean by hard in this context.

These questions do not have complete answers at this moment. Research into this area has just begun. We hope that this text will provide some stimulus for inquiries into such types of questions.

Problems

1.1 Prove that a^k has complexity $O(k)$ for both serial and parallel computers.

1.2 Can you find another example where the serial and the parallel complexity are the same?

1.3 One application where we can indeed generate the set of states in parallel and do not parallelize the mapping is the **simple sampling**. In simple sampling there is no dependence of the newly generated state on the previous state. An example of simple sampling is the random walk problem. Can you think of more such examples?

1.4 Suppose that the set of states which we want to generate are random numbers. Would this be an example where we could use parallelism for the generation of the set?

2. Computer Simulation Methods

In this chapter we give a condensed introduction to the basic features of computer simulation methods. We particularly review the Monte Carlo method of importance sampling and the principles of molecular dynamics simulations. Some of the systematic effects associated with computer simulations, such as finite-size effects, statistical errors and so on, are reviewed. The Ising model is introduced in order to provide a concrete illustration of the Monte Carlo method, which will be useful also in later chapters as a simple context in which to present some of the algorithms we will discuss. It is assumed that the reader has at least some prior understanding or experience of the material covered in this chapter, since it is meant mainly as a reference base for the material addressed in subsequent chapters. For a more complete description of the material presented here we recommend referring to one of the number of books that are now available [2.1–5]. The reader embarking upon computer simulations for the first time is encouraged to consult one of these.

2.1 Essential Features of Simulation Methods

Computer simulations have progressed over the past couple of years to become an indispensable tool for investigations in many branches of physics and of the physical sciences more generally. Such simulations have enabled the solution of a wide range of problems that would otherwise have remained inaccessible, and have also led to new insights and new lines of enquiry. This progress has, of course, been closely associated with developments in computer technology, which have brought about many-fold increases in computer speed, memory size and flexibility. With the recent appearance and growing widespread use of parallel computers many new possibilities have opened up, which we shall present and discuss in this book.

Before doing so, however, it is necessary to review the general features of computer simulations in the context of statistical mechanics. Our aim here is to provide an overview of the techniques most widely used, and where appropriate to illustrate these by examples.

2.1.1 Ensemble Averages on a Computer

A statistical physics system is described by a set of variables whose values contain all the possible *degrees of freedom* of the system. A magnetic system, for example, can be described by a set of *spins* $\{s_i\}$ which live on each site i of a lattice. A particular set of values for the spins is a *configuration* (corresponding to a microstate) and the *phase space* Ω is the set of all possible spin configurations of the system. We will be using various models as illustrations – particularly the *Ising model*, which will be discussed more fully in Sect. 2.6.1. The energy function [or *Hamiltonian* $\mathcal{H}(s)$, where $s = \{s_i\}$] of a system describes the energy in terms of interactions between the degrees of freedom. Each configuration has associated with it a probability, the *Boltzmann factor*, given by

$$P(s) = \frac{1}{Z} \exp[-\mathcal{H}(s)/k_\mathrm{B}T] \quad , \tag{2.1}$$

where the *partition function* Z is a normalization factor:

$$Z = \sum_{\{s\}} \exp[-\mathcal{H}(s)/k_\mathrm{B}T] \tag{2.2}$$

with the sum going over all phase space (i.e., all possible configurations). T is the absolute temperature and k_B is the Boltzmann constant. The Boltzmann factor can be thought of as a probability density which describes the statistical weight of the configuration s occurring in thermal equilibrium.

Typically we are interested in computing various thermodynamic quantities, which are given (here in the canonical ensemble) by weighted averages:

$$\langle A(s) \rangle = \frac{1}{Z} \sum_{\{s\}} A(s) \exp[-\mathcal{H}(s)/k_\mathrm{B}T] \quad . \tag{2.3}$$

This enables us to compute large-scale (macroscopic) quantities, such as average energy and magnetization:

$$E = \langle \mathcal{H} \rangle / N \quad , \qquad M = \langle \sum_i s_i \rangle / N \tag{2.4}$$

(for a system with N spins) as the average over the (microscopic) degrees of freedom of the system.

2.1.2 Simulation Algorithms

In a computer simulation of a statistical mechanical system we are able to perform a *numerical experiment* which differs from its laboratory counterpart in a number of crucial ways. Firstly, of course, we are able to choose and vary our model as we wish, thus enabling us to investigate the *basic assumptions* of our understanding of the systems we investigate. It is also possible to choose the parameters of the model, such as the temperature, interparticle coupling strengths and so on, to have any value we desire. This gives us both an enormous range of phenomena that can be investigated and a great flexibility in terms of models and conditions that we can simulate. The limitations of this program are very different from those encountered in the laboratory. In later sections we will discuss some of these limitations, such as the finite simulation time and resulting limitation on the accuracy of the results, the effect of the finite size of the systems investigated and slow relaxational modes.

In the sections that follow we discuss two of the principal methods used in computer simulations, namely the Monte Carlo algorithm and molecular dynamics methods. In the Monte Carlo algorithm a sequence of successive configurations are generated with the help of (pseudo)random numbers and a suitably chosen transition probability (such as that given by the Metropolis algorithm [2.6], while in the molecular dynamics algorithms the classical equations of motion are numerically integrated forward in time. These represent two of the most important methods in a whole spectrum of possible algorithms, and by viewing the Monte Carlo process dynamically we shall see that the two methods are in fact closely related.

It is not possible here to discuss the multitude of other methods and their many variations so we limit ourselves to some brief remarks and references, which the interested reader is encouraged to pursue. In Sect. 2.6.1 we introduce and discuss the Ising spin model [2.7] and a number of the algorithms we introduce are illustrated by applying them to the Ising model. Other systems, including both spin models and gauge systems, are introduced either in the text or in the exercises throughout the book, but always with appropriate explanation and references to the literature. In the outline of the molecular dynamics method in Sect. 2.3 we consider only simulations in the microcanonical ensemble, in which the total energy of the system remains constant, and where the particle interactions are given by two-particle forces. It is possible also to introduce random forces and thus investigate systems in the canonical ensemble, which may be useful in studying problems like Brownian motion. The molecular dynamics method can also be applied to spin or lattice gauge models [2.8] by introducing a spurious time variable that is identified with the computer run-time. By incorporating random fluctuating forces appropriately it is possible to show that the resulting Langevin equation [2.9] generates spin configurations with the correct Boltzmann probability distribution. The hybrid algorithm [2.10,11] provides a smooth interpolation between the Langevin algorithm and molecular dynamics algorithms in which a parameter can be adjusted to optimize the frequency with which the velocities are refreshed (i.e., chosen randomly according to the appropriate Boltzmann probability distribution). This method can further be combined with the Metropolis algorithm to form the hybrid Monte Carlo algorithm [2.12] where the hybrid method is used to provide a trial configuration that is accepted or rejected according to the Metropolis criteria (2.14), thereby eliminating the finite-step-size effects associated with molecular-dynamics-type algorithms. Another update algorithm suitable for spin models like the Ising model is the cellular

automata algorithm Q2R [2.13–15], in which only spins with an equal number
of up and down neighbours are flipped. This algorithm conserves the energy of
the system, and it is therefore necessary to explicitly choose the initial configuration with the required energy. In the demon algorithm of *Creutz* [2.16] an extra
degree of freedom (the demon) is introduced which travels around the system
transferring energy from one site to another.

Recently there has been considerable progress made in developing algorithms
to combat the problem of critical slowing down, a problem that we will discuss
further in Sect. 2.5.2. Such methods include overrelaxation algorithms [2.17,18],
Fourier acceleration [2.19], multigrid [2.20–22] and cluster algorithms [2.23,24].
In these algorithms the underlying dynamics is modified in such a way that the
low-frequency modes of the system, which control the long-distance behaviour,
are speeded up. In Sect. 2.6.5 we shall look at cluster algorithms more closely
and details of their implementation are discussed in Sect. 2.6.4. Details of parallel
implementations are discussed in Chap. 7.

2.2 The Monte Carlo Algorithm

The Monte Carlo simulation of a statistical mechanics model with a Hamiltonian
function $\mathcal{H}$ enables us to construct a set of configurations which are distributed
with the appropriate probability. This ordered sequence of configurations is generated using (pseudo)random numbers (the importance of a good random number
generator cannot be overstressed – a good discussion of the issues involved is
found in [2.3] and a generator is presented here in Appendix B.). The resulting measurements of physical quantities over the configurations corresponds to
a "time" averaging in which the system can be thought of as evolving according
to some stochastic kinetics. This dynamical interpretation of the Monte Carlo
process, which is further discussed in Sect. 2.2.3, enables us also to use Monte
Carlo methods to investigate transport and relaxation behaviour of systems in
(physical) time.

2.2.1 Simple Sampling

On attempting to calculate the thermodynamic quantities associated with a statistical mechanics system, as given by (2.3), on a lattice of any reasonable size,
it immediately becomes clear that summing over *all* possible configurations is
not possible. The most straightforward and naive way to proceed is then to try
approximating the summation over all phase space Ω by choosing a statistical
sample of points in phase space $\{s_1, s_2, ..., s_M\}$, which is analogous to approximating an integral by a discrete sum. Each sample configuration is weighted by
its Boltzmann factor $P(s)$, as given by (2.1). The discrete sum

$$\overline{A(s)} = \frac{\sum_{i=1}^{M} \exp[-\mathcal{H}(s_i)/k_B T] A(s_i)}{\sum_{i=1}^{M} \exp[-\mathcal{H}(s_i)/k_B T]} \qquad (2.5)$$

then clearly provides a good approximation to the expectation value (2.3) in the limit $M \to \infty$. Typically, of course, the actual number M of configurations summed represents only a tiny fraction of the phase space, and these configurations are a random set of points in phase space that are chosen using some pseudorandom number generator (configurations chosen according to a regular pattern can lead to systematic errors, but we won't go into that here).

The advantage of this method is that each configuration is independent of all others, and therefore we can use all the usual statistical methods to evaluate the accuracy of the results obtained. We very quickly find, however, that this is often a very inefficient means of calculating thermodynamic quantities, because most of the configurations give a negligible contribution to the sum in (2.5). This can be seen straightforwardly by considering the case where $A(s)$ in (2.5) is the Hamiltonian $\mathcal{H}(s)$ itself. For a system with N degrees of freedom it is straightforward to show that, for N sufficiently large, the relative fluctuation goes as

$$\frac{\langle \mathcal{H}^2 \rangle - \langle \mathcal{H} \rangle^2}{\langle \mathcal{H} \rangle^2} \propto \frac{1}{N} \quad . \qquad (2.6)$$

The probability distribution $P(E)$ of the energy E per degree of freedom, defined by

$$P(E) = \frac{1}{Z} \sum_{\{s\}} \delta_{\mathcal{H}(s),NE} \exp[-\mathcal{H}(s)/k_B T] \quad , \qquad (2.7)$$

is very sharply peaked, with a peak height of $\sqrt{N}$ and a width of $1/\sqrt{N}$. Consequently, most of the states generated by a simple sampling technique will give a negligible contribution.

2.2.2 Importance Sampling

A more efficient technique is not to sample at random but rather to sample configurations from the part of phase space that is important at the temperature T. If a configuration s_i is chosen with a probability $P(s_i)$ then the thermodynamic average becomes

$$\overline{A(s)} = \frac{\sum_{i=1}^{M} \exp[-\mathcal{H}(s_i)/k_B T] A(s_i)/P(s_i)}{\sum_{i=1}^{M} \exp[-\mathcal{H}(s_i)/k_B T]/P(s_i)} \quad . \qquad (2.8)$$

It is clear that if we choose the probability $P(s_i)$ to exactly cancel the Boltzmann factor then this reduces straightforwardly to the simple arithmetic average of the contributions:

$$\overline{A(s)} = \frac{1}{M} \sum_{i=1}^{M} A(s_i) \quad . \tag{2.9}$$

This technique enables us to evaluate thermodynamic quantities by averaging only over a sample of configurations that give a significant contribution, and it is thus called *importance sampling*.

A simple method for implementing this procedure is the *Metropolis* algorithm [2.6], in which the successive configurations are not chosen independently of each other but rather are constructed via a *Markov chain*. The successive states are generated with a particular transition probability $W(s_i \rightarrow s_{i+1})$, which is chosen so that in the limit of a large number of configurations M the probability $P(s_i)$ tends towards the equilibrium distribution

$$P_{eq}(s_i) = \frac{1}{Z} \exp\left[-\mathcal{H}(s_i)/k_B T\right] \quad . \tag{2.10}$$

A sufficient condition to ensure this is the principle of detailed balance:

$$P_{eq}(s)W(s \rightarrow s') = P_{eq}(s')W(s' \rightarrow s) \quad . \tag{2.11}$$

We see immediately that the ratio of the transition probabilities depends only upon the change in energy:

$$\frac{W(s \rightarrow s')}{W(s' \rightarrow s)} = \exp\left(\frac{\mathcal{H}(s) - \mathcal{H}(s')}{k_B T}\right) \quad . \tag{2.12}$$

This does not, however, uniquely specify the form of the transition probability W. This remaining freedom in the choice of W enables us to choose various schemes, of which two of the most widely used are given by [2.6,25]

$$W(s \rightarrow s') = \begin{cases} \frac{1}{\tau_s}\exp(-\Delta\mathcal{H}/k_B T) & \text{if } \Delta\mathcal{H} > 0, \\ \frac{1}{\tau_s} & \text{otherwise,} \end{cases} \tag{2.13}$$

$$W(s \rightarrow s') = \frac{1}{2\tau_s}\left[1 - \tanh\left(\frac{\Delta\mathcal{H}}{2k_B T}\right)\right] , \tag{2.14}$$

where τ_s is here an arbitrary factor that we can set equal to unity. The reader interested in more technical details, such as proving that the above choices for the transition probability W actually give a probability distribution $P(s)$ that converges towards $P_{eq}(s)$, is encouraged to consult some of the more detailed literature (see [2.1–5] and references therein).

The *inherent* parallelism contained in this update procedure will play an important role in our subsequent discussion of how to construct efficient parallel algorithms for Monte Carlo simulations. The fact that $\Delta\mathcal{H}$ typically contains only very local interactions means that many variables can be updated simultaneously without any conflict with the requirement of detailed balance.

2.2.3 Interpretation of the Monte Carlo Process as a Dynamical Process

The Monte Carlo algorithm may be interpreted via a master equation as describing a dynamic model with stochastic kinetics. Recall that we introduced a factor τ_s in (2.13,14), which we can now use to normalize the "time" scale associated with the generation of successive configurations, the time unit becoming one Monte Carlo step per degree of freedom. The probability $P(s, t)$ becomes time dependent and gives the probability that at time t the Monte Carlo process generates a configuration s. It satisfies the Markovian master equation

$$\frac{P(s, t)}{dt} = \sum_{s'} P(s', t) W(s' \to s) - \sum_{s'} P(s, t) W(s \to s') \quad , \tag{2.15}$$

which describes the flux flow of the probability of states entering (the first term) and leaving (the second term) s.

This dynamic interpretation is particularly useful for the Ising model, since the commutator $[s_i, \mathcal{H}_{\text{Ising}}]$ vanishes and the Ising model therefore has no *intrinsic* dynamics (unlike, for example, the Heisenberg ferromagnet where the commutator does not vanish). Thus we can interpret the Monte Carlo process as simulating a spin system that is weakly coupled to a heat bath, which induces random spin flips. This enables the investigation of such processes as relaxation in glasses [2.26], quenching [2.27,28], nucleation [2.27] and spinodal decomposition [2.29,30]. There are many other applications where this approach is also useful, such as the study of the Brownian motion of macromolecules [2.31,32] and diffusion processes [2.33,34].

The equilibrium properties of a system are typically given by the long time properties, in which the probability distribution $P(s, t)$ is in a steady state and no longer varies with time. This condition, namely $dP(s, t)/dt = 0$, is exactly that of detailed balance (2.11) and gives the Boltzmann probability distribution $P(s, t) = P_{\text{eq}}(s)$. The averages of thermodynamic quantities which we measure (2.9) are then simply time averages along the trajectory in phase space, which is discretized into finite time steps corresponding to successive configurations separated by one Monte Carlo update, which is the natural unit of time. The trajectory itself is controlled by the master equation (2.15) and reflects the stochastic nature of the Monte Carlo process.

In practice this provides a means of constructing a very useful guide to whether a system has reached equilibrium and to how correlated successive configurations are. In the early stages of a simulation the initial configuration often plays an important role, and the relaxation of the system into equilibrium can be monitored by plotting the time dependence of the measured physical variables. This is a widely used means of establishing the relaxation time τ_R in which a system comes into equilibrium, especially when very different starting configurations are compared. A note of warning is however necessary here, since different variables will not in general relax into equilibrium at the same time – indeed, long-range quantities, such as those associated with long-distance correlations, typically require much longer than local quantities (such as energy density and magnetization) to reach equilibrium.

The time correlation of the physical variables A and B between equilibrium configurations separated by time τ is

$$\overline{A(\tau)B(0)} = \frac{1}{t_M - \tau - t_0} \int_{t_0}^{t_M - \tau} A(t' + \tau)B(t')dt' \quad , \quad t_M - \tau > t_0 , \quad (2.16)$$

where t_0 is typically the time taken for the system to reach equilibrium. The normalized correlation function, given by

$$C_{AB}(\tau) = \frac{\overline{A(\tau)B(0)}}{\overline{A(0)B(0)}} \quad , \quad (2.17)$$

typically has an exponential behaviour $C(\tau) \propto e^{-\tau/\tau_c}$ (for Monte Carlo algorithms involving only local updates of the variables), where τ_c then provides a measure of the time correlation between successive configurations.

The normalized autocorrelation function of the quantity A is given by

$$\varrho_A(t) = \frac{\langle A(0)A(t) \rangle - \langle A \rangle^2}{\langle A^2 \rangle - \langle A \rangle^2} \quad . \quad (2.18)$$

We can define two autocorrelation times: the integrated one

$$\tau_{\text{int},A} = \frac{1}{2} \sum_{t=-\infty}^{\infty} \varrho_A(t) \quad (2.19)$$

and the exponential one

$$\tau_{\text{exp},A} = \lim_{t \to \infty} \sup \frac{t}{-\log \varrho_A(t)} \quad . \quad (2.20)$$

Usually the two autocorrelation times are the same, although there are some models, such as the self-avoiding random walk [2.35], where they are different. An estimate for the error $\Sigma_{A,N}$ of the measurement of the quantity A in a Monte Carlo simulation with N measurements (which are correlated) is given by (see Ref. [2.2, p.34])

$$\Sigma_{A,N}^2 = \frac{(\langle A^2 \rangle - \langle A \rangle^2)2\tau_{\text{int},A}}{N} \quad , \quad (2.21)$$

with $\tau_{\text{int},A} \gg 1$ given by (2.19) above. In practice the sum in (2.19) will be swamped by noisy contributions from large separations and it is necessary to introduce a truncation window [2.35,36]

$$\tau_A(W) = \frac{1}{2} + \sum_{t=1}^{W-1} \varrho_A(t) + R(W) \quad (2.22)$$

with the remainder

$$R(W) = \varrho(W)\frac{1}{1 - \kappa(W)} \quad , \quad \kappa(W) = \frac{\varrho(W)}{\varrho(W-1)} \quad . \tag{2.23}$$

Typically $\tau_A(W)$ will be a smooth monotonically increasing function of W which can be straightforwardly extrapolated.

This relationship between time averages and the canonical ensemble average leads to the question of ergodicity, which needs careful attention. This is, in general, a very difficult problem that must be analysed for each model. Suffice it here to make some general remarks. Firstly, if we have only finite potentials in our model, simulations on finite lattices will necessarily be ergodic. However, practical problems can arise if our time trajectory (i.e., length of simulation time) is not sufficiently long, i.e., shorter than the so-called "ergodic time" τ_e [2.26,37,38]. This is intimately connected with the finite-size of the lattice, and we will discuss this further when we consider the relationship between finite-size effects and time correlations in Sect. 2.5. More fundamental ergodicity problems may arise if our model contains infinite potentials, such as occur in the self-avoiding random walk problem [2.35], where certain sets of configurations may (depending on the details of the algorithm) be mutually inaccessible. The reader who is interested in simulating such models is well advised to consult the literature on this point [2.39].

2.3 Molecular Dynamics

In the molecular dynamics method the phase space trajectories for a collection of molecules are computed using the classical laws of motion. Since the first molecular dynamics simulations of a hard sphere fluid [2.40] more than three decades ago, this field has developed dramatically. A wide range of phenomena, including molecular systems, Brownian motion of macromolecules, polymers and quantum systems, have all been investigated, each requiring appropriate molecular dynamics algorithms. A variety of statistical ensembles can also be simulated, which require appropriate modifications of the basic formalism [2.41]. A number of excellent reviews [2.3,42–47] are available which cover many of the various facets of molecular dynamics in some detail. We restrict ourselves here to an overview of the basic ideas and methods for the dynamics of atomic systems in the microcanonical ensemble, although these methods can be extended to the canonical ensemble and other ensembles [2.41]. For a discussion of molecular dynamics methods appropriate to the canonical ensemble, Langevin equation, polymer materials and quantum systems the reader is directed to the above references.

The molecular dynamics method requires a well-defined microscopic description of a physical system, which may be given in terms of a Hamiltonian, a Lagrangian or directly by Newtons equations of motion. The thermodynamic quantities are obtained by simulating the dynamics of the microscopic constituents. This differs from the Monte Carlo method, where we are interested primarily in obtaining the configurational properties of the system, although it is possible to give the Monte Carlo method a dynamic interpretation in terms of the system being coupled to a heat bath at the appropriate temperature (recall Sect. 2.2.3). This stochastic dynamics which is imposed upon the system by the Monte Carlo method does not generally correspond to the intrinsic dynamics of the system. The molecular dynamics method, on the other hand, computes the actual phase space trajectories for a collection of molecules. The accuracy of the method is governed principally by the particular algorithm used, the total length of the trajectory in phase space and by the discretization of the particle paths, which in turn will be influenced by the speed and memory of the computer.

2.3.1 The Microcanonical Ensemble

In the microcanonical ensemble a system evolves along a trajectory in phase space in which the energy E, the number of particles N and the volume all remain constant. The Hamiltonian describing classical pair-wise interaction of N particles is given by the sum of the kinetic and potential energy contributions

$$\mathcal{H} = \frac{1}{2} \sum_i \frac{p_i^2}{m_i} + \sum_{i<j} V(r_{ij}) \quad , \tag{2.24}$$

where p_i is the momentum of particle i and r_{ij} is the distance between particles i and j. In order not to unnecessarily complicate the exposition here we have neglected orientation-dependent interactions and consider monatomic systems, although the methods are, of course, more generally applicable. The temperature of the system is given by the mean kinetic energy $\langle K \rangle$ via the equipartition theorem:

$$K = \frac{1}{2} \sum_{i=1}^{N} m_i v_i^2 = \frac{1}{2}(f - 3)k_B T \quad , \tag{2.25}$$

where $f = 3N$ is the number of degrees of freedom for N atoms, and three degrees of freedom have been removed by constraining the centre of mass of the system to be at rest. The temperature is clearly not a constant of the simulation and can fluctuate. The angular momentum of the system is usually not constrained to zero, since it is not conserved for systems with periodic boundary conditions.

The classical equations of motion for the system are

$$m \frac{d^2 r_i(t)}{dt^2} = \sum_{i<j} F_i(r_{ij}) \quad , \tag{2.26}$$

where F_i, the force on atom i, is given by the gradient of the potential

$$F_i = - \sum_{j \neq i}^{N} \nabla_{r_i} V(r_{ij}) \quad . \tag{2.27}$$

As in the analytic solution of a second-order differential equation, we need to specify both the initial positions r_i and velocities v_i of the atoms. For the simulation of solids it is natural to place the atoms at their equilibrium lattice positions. A liquid configuration can be initialized in the same way, and the lattice parameter then expanded to give the required density. The liquid is then "heated" to "melt" the crystal structure by raising the initial temperature and subsequently scaling it down to the required value. The initial velocities are chosen randomly and scaled to the desired temperature according to (2.25) by scaling each of the velocities by an appropriate correction factor [2.48], which is done during the equilibration phase of a simulation. Usually the initial velocities are chosen from a Maxwell-Boltzmann distribution, although this is not necessary because the system will rapidly be driven into this distribution and lose all memory of its initial state. When the system has equilibrated, the mean values of the kinetic and potential energy will have settled, although in practice it sometimes requires care to distinguish long-lived metastable states from equilibrium, especially near two-phase coexistence. It may also happen that the initial state of the system is such that only an irrelevant part of phase space is investigated within the finite time of the simulation, and it is therefore necessary to perform simulations from different initial conditions.

The ensemble average of a quantity A is given by the time average over the configurations generated as

$$\langle A \rangle = \bar{A} = \lim_{t' \to \infty} \frac{1}{(t' - t_0)} \int_{t_0}^{t'} dt\, A(t) \quad , \tag{2.28}$$

where we use the notation $\langle A \rangle$ to denote ensemble averages and $\bar{A}$ for trajectory averages. t_0 is the equilibration time and $(t' - t_0)$ is the time over which the trajectory is averaged, which in a typical simulation corresponds to a time scale up to some tens of nanoseconds. The time t_0 must be sufficient to establish the equipartition of kinetic energy between the degrees of freedom as well as the correct distribution of the total fixed energy between the potential and kinetic energies. Ergodicity ensures that the time average over a sufficiently long trajectory is equivalent to the ensemble average. However, in a computer simulation some care is required to ensure that the total length of the simulation is sufficient to explore phase space adequately, which in general depends upon the properties of the system being investigated.

2.3.2 Discretization and Systematic Effects

The differential equations of motion for the molecules are solved by constructing a finite difference scheme with step size h, which together with the number of steps iterated determines how much of phase space will be sampled. It is clear that we would like to choose h to be as large as possible in order to sample as much of the phase space as is feasible. However, we need to consider both the time scale on which changes in the system occur and the error associated with the discretization, which will depend upon the algorithm used. Typically h lies in the range 10^{-16}s to approximately 10^{-14}s, depending upon the particular potential being considered, and the actual value of h chosen requires careful judgement. Two types of algorithms suitable for such problems are predictor-corrector algorithms [2.49] and Verlet-type algorithms [2.50]. The most widely used of these are the Verlet-type algorithms, which we outline below and which allow a slightly larger time step h and require less storage.

The basic Verlet algorithm is given by

$$r(t + h) = 2r(t) - r(t - h) + h^2 a(t) \quad . \tag{2.29}$$

The acceleration is given straightforwardly by the force $a_i(t) = F_i(t)/m_i$. A simple improvement of this scheme is the leap-frog algorithm, [2.51], in which the velocities are defined at the half time steps:

$$v\left(t + \frac{h}{2}\right) = v\left(t - \frac{h}{2}\right) + ha(t) \quad ,$$

$$r(t + h) = r(t) + hv\left(t + \frac{h}{2}\right) \quad . \tag{2.30}$$

The positions and velocities are then iterated alternately in a leap-frog fashion. A further improvement is given by the velocity-Verlet algorithm [2.52]

$$r(t + h) = r(t) + hv(t) + \frac{h^2}{2}a(t) \quad ,$$

$$v\left(t + \frac{h}{2}\right) = v(t) + \frac{h}{2}a(t) \quad ,$$

$$v(t + h) = v\left(t + \frac{h}{2}\right) + \frac{h}{2}a(t + h) \quad , \tag{2.31}$$

which has a close resemblance to a three-value predictor-corrector algorithm. In this form of the Verlet algorithm the numerical stability is enhanced. A discussion of the finite-step-size error associated with the various discretization methods is given in [2.53].

In order to minimize the problem of the finite size of the sample and unwanted surface effects we use periodic boundary conditions (Sect. 2.5.1). We can then think of our box (or *cell*) of linear dimension L as being surrounded in each direction by an infinite sequence of identical boxes, each containing the same particles in the same relative positions. The question then arises as to whether we sum the contributions from the forces exerted by all the copies of a molecule or whether we consider only the nearest. For long-range potentials, such as for Coulomb or dipole-dipole interactions, the interaction with all the images can be elegantly included using the Ewald method [2.54–56]. In most other cases it is sufficient to use the minimum image convention, in which each molecule interacts only with the *nearest* image of the other molecules. We can think of each molecule as being at the centre of its own box of linear dimension L and only interacting directly with other molecules within this box, which provides a good approximation provided the forces are negligible beyond a range of $L/2$. In this case a spherical cutoff of the potential at a distance less than $L/2$ is frequently introduced, which also reduces the number of force calculations that are necessary. There are situations, however, where the effect of a sharp cutoff of the potential requires particular attention, such as the relaxation of a non-equilibrium state into equilibrium [2.57]. As long as the volume is large enough, the actual shape of the molecular dynamics cell is irrelevant for simulations of gases and liquids. However, the shape of the cell plays an important role for solid structures in a crystalline state, since the lattice vector of the solid must be commensurate with the cell. Usually a solid stays in its initial configuration, although this may not necessarily correspond to a minimum of the free energy, unless the cell is explicitly allowed to change shape [2.58].

The range of the intermolecular potential clearly plays a crucial role in the amount of computation required in any simulation. A discussion of the various algorithms suitable for systems with short-, medium- and long-range interactions is postponed until a later chapter when we consider the parallelization of molecular dynamics algorithms.

2.3.3 Molecular Dynamics Algorithms

There are two basic techniques for carrying out molecular dynamics simulations in which the interparticle potential has some effective cutoff, i.e., a distance r_{co} beyond which we do not need to explicitly calculate the forces between two particles, although such forces may be included by some averaging technique [2.59]. The first method, due to *Verlet* [2.60], involves constructing a particle neighbour table to keep a record of which particles are in the vicinity of a particular particle. An alternative method, called the link-cell method [2.61], proceeds by dividing the volume into subcells and using linked lists within these subcells to facilitate the identification of pairs of molecules that are close enough to interact.

In the neighbour table method each of the $N(N-1)/2$ distances is calculated only every n^{th} step, and for each particle a table is constructed of all the particles that are within a distance r_M of that particle. The distance r_M is chosen to be sufficiently large that no particle can move into the range r_{co} of the potential within the next n steps (clearly $r_M > r_{co}$). Only the particles within the shell between r_M and r_{co} can start to interact during these n steps. This method clearly saves a very substantial amount of computer time for large systems, since the number of particles within the shell is very much less than the total number of particles. A number of improvements and variations upon this method are possible, such as introducing a whole series of larger and larger shells that must only be updated at correspondingly longer time intervals.

The link-cell method proceeds by dividing the molecular dynamics cell into a number of equally sized subcells with widths greater than r_{co}. At every time step a linked list is created of the particles within each subcell, whereby each particle has a pointer associated with it that identifies another particle in the subcell (or indicates that it is the last in the list). In order to calculate the force on any particular particle it is only necessary to consider particles within the same and immediately adjacent subcells.

The various advantages and limitations of these methods have been compared [2.62] for calculations on scalar computers. In later chapters we will discuss parallel algorithms [2.63,64] that are suitable for molecular dynamics simulations. Which particular algorithm is suitable in any particular situation will depend upon the characteristics of the computer and the physics of the model, and particularly the range of the interparticle potential and density of particles.

2.4 Hybrid Molecular Dynamics

One of the problems with the molecular dynamics method (which can also in some circumstances be an advantage) is that temperature enters through the back door. The formulation of molecular dynamics simulation algorithms [2.3,45,46], whether they are based on Newton's equations, the Lagrangian formulation or start from Hamilton's equations, are all for simulations at a constant energy. The method by which temperature is typically introduced into the formulation is to scale the velocities of the particles, and in one way or another almost all schemes come down to this. The only exceptions are schemes where some form of stochasticity is involved [2.41,65,66].

In Monte Carlo schemes, on the other hand, the temperature T is an integral part of the basic algorithm, whereas energy does not usually remain constant, although there are such algorithms [2.16,67]. Monte Carlo algorithms typically suffer a drastic reduction in their updating efficiency if we consider a global update of the configuration. Usually we proceed by proposing a *local* move for one particle which is accepted or rejected according to a particular update probability. Should we try to propose a move to the Monte Carlo step where all particles are given the chance to change their positions at the same time, then the move is almost certainly rejected.

However, this is not so for molecular dynamics, where the positions of all particles are updated at the same time without such conflicts. The question arises whether it is possible to design an algorithm with a *global* update of the configuration of a system and at the same time a constant temperature [2.68].

Consider again a system of N particles inside a box of linear dimension L. The Hamiltonian with kinetic (K) and potential energy (U) parts is

$$\mathcal{H} = K + U \tag{2.32}$$

$$= \frac{1}{2}\sum_i mv_i^2 + \sum_{i<j} V(r_{ij}) \quad , \tag{2.33}$$

where m and v_i denote the mass and the velocities respectively, and r_{ij} is the distance between two particles i and j at $\boldsymbol{r}_i$ and $\boldsymbol{r}_j$.

The particle positions inside of the box are determined by the integration of the equations of motion resulting from the Newtonian equations of motion (2.26,27)

$$m\frac{dv_i}{dt} = -\sum_{j \neq i}^{N} \nabla_{r_i} V(r_{ij}) \quad , \quad i = 1, ..., N \quad . \tag{2.34}$$

The velocities of the particles are determined by the forces acting on them and are related to the temperature by the equipartition theorem [2.69], as given in (2.25). Because of this relation it is possible to scale the velocities with a parameter α related to the desired temperature

$$v_i \rightarrow \alpha v_i \quad , \tag{2.35}$$

such that the overall system has the correct temperature.

From the statistical mechanics point of view this is perfectly all right since calculating the configurational part

$$Z_U = \sum_r \exp(-U/k_B T) \tag{2.36}$$

of the partition function

$$Z = \sum_{r,v} \exp(-\mathcal{H}/k_B T) \tag{2.37}$$

is the difficult step. The momentum part can always be integrated out. Having done the step of scaling the velocities (i.e., manually rearranging the system) we ask if we can go a step further and discard all velocities and replenish them from a distribution at exactly the required temperature.

We want to interpret the molecular dynamics method not so much as an integration of the equations of motion, but as a Markov chain. In a Monte Carlo algorithm for the evaluation of the many-body system described by the Hamiltonian (2.33), the configurational part of the partition function is sampled such that the result is a Markov chain of configurations. The Markov chain is constructed such that the transition between states reflects the temperature and the equilibrium distribution of configurations. The transition between states in our example of particles in a box is done by randomly displacing a particle and accepting such a displacement with the correct probability.

The central point is that we can interpret the molecular dynamics algorithm as really generating such a chain. The difference to the canonical formulation of the equations is, of course, that energy is conserved. Instead of always generating the transition from one configuration to the next with the help of the integration resulting from the velocities and the forces, we introduce a Metropolis procedure to accept or reject such proposed configurations. This differs from the hybrid Monte Carlo method [2.12,67] proposed in the context of lattice gauge theories, where the entire change in the Hamiltonian is considered in the update probability of the configuration.

Let r be a configuration of the particles and let Δ be a discretization scheme for the integration of the equations of motion. The new positions of the particles r' and velocities are calculated as

$$r' = \Delta(r, v, f, h) \quad , \tag{2.38}$$

where v and f denote the velocities and the forces from the previous step, and h is the step size in the time direction.

After one step of applying Δ we have gone from a configuration r to a new one r'. Such a move changes the potential energy U by $\delta U = U(r') - U(r)$. Of course, in the integration scheme the kinetic energy is also affected, but we neglect this change since we disregard the kinetic energy part in the Hamiltonian and only use the velocities as propagators.

A Metropolis acceptance procedure is now applied to the potential energy change δU and we accept the new configuration with a probability

$$P^{\text{hybrid}} = \min\{1, \exp(-\delta U/k_B T)\} \quad . \tag{2.39}$$

If the configuration is accepted we continue with the velocities as they are, i.e., those resulting from the application of the operator Δ. If the configuration is rejected, we draw a new set of velocities v' from a Maxwell distribution at the temperature T and apply Δ once more:

$$r' = \Delta(r, v', f, h) \quad . \tag{2.40}$$

The transformation from v to v' will be denoted by $\mathcal{T}$.

We have constructed a global transition probability which leads from a set of coordinates r to a new set of coordinates r' for the particles

$$W^{\Delta}_{T,h}(r \to r') \quad . \tag{2.41}$$

What is the major difference between a usual Monte Carlo algorithm and the algorithm described above? Whereas $W(r \to r')$ is *local* in the usual Monte Carlo algorithm, it is *global* here since it affects the entire system. In the Monte Carlo algorithms only a single atom is displaced, because a displacement of the entire set of atoms would result in a very low acceptance rate. With the hybrid algorithm we succeed in incorporating collective modes not present in the local Monte Carlo case.

The transition probability is by construction positive. Do we fulfil detailed balance or, even better, microscopical reversibility, which is necessary to guarantee that ultimately the states we generate are distributed according to the correct distribution at the temperature T? We can see that this is the case because if we choose a formulation of the integration procedure which is time reversible, then by choosing the appropriate velocities we can definitely reverse the step and return to the previous set of coordinates!

Hence we generate a trajectory in configurational space

$$r_0, r_1, ..., r_{n-1} \tag{2.42}$$

which is no longer correlated in the sense of the microcanonical ensemble molecular dynamics but rather according to the canonical ensemble [2.70,71].

2.5 Accuracy Considerations and Finite-Size Problems

In the previous sections we discussed how to go about performing a Monte Carlo simulation. This involves, of course, initially deciding how large the system we wish to simulate should be (i.e., how many variables N, such as spins or particles, should be included) and how much time is required in order to generate configurations and measure particular quantities. We then need to consider how reliable our measurements are and how they can be used to gain information about the thermodynamic limit $N \to \infty$, since it is usually the bulk properties of the infinite system that we are interested in. It is clearly not possible for a finite system with a non-singular Hamiltonian to exhibit singular behaviour, and such systems therefore do not have true phase transitions. Nevertheless, it is possible to gain substantial information about singular phenomena in infinite systems, such as phase transitions, by systematically studying the dependence of the quantities of interest on the size of finite systems.

These questions are all intimately related. The size of system, for example, plays a large role in determining the time it takes for the configurations to reach equilibrium, their subsequent correlation, and the statistical accuracy of the results. The approximations that are necessary are, however, controllable and their influence on the results can be determined by systematically studying the effects of finite size and finite simulation time. There are no fixed rules for the judgements that must be made in these matters, which often depend upon the physics of the problem being studied, but our aim in this section is to provide general guidelines and outline the most important features. We will only discuss effects here that are common to all simulations on scalar computers. Special considerations that may be necessary when the algorithms are partitioned over a number of parallel processors will be discussed in the relevant sections in later chapters.

Finite-size effects are not merely a computational source of error that must somehow be eliminated – they can provide us, with the help of finite-size scaling [2.72], with a great deal of information about the thermodynamic quantities of interest. Finite-size scaling has become a powerful tool in computer simulation studies and enables us to extract infinite system size results precisely by studying the size dependence of results on relatively small volumes. In the following sections we first look briefly at the most important finite-size effects, and then in Sect. 2.5.4 we outline the central features of finite-size scaling.

2.5.1 Choosing the Boundary Conditions

On a system of finite size it is only possible to accommodate structures whose length scales do not exceed the linear dimension of the simulated system. If we are interested, for example, in studying a system near a critical point T_c at which the correlation length ξ diverges, then the finite size dependence of the data needs to be carefully investigated. Since we are usually interested in the bulk properties of the infinite system we need also to consider the effect that the boundary conditions have upon the measurements. There are a number of situations where surface effects can be usefully employed to study problems involving surfaces and interfaces, but we will not consider these here.

Perhaps the most widely used boundary conditions on regular lattice systems and systems confined to a box (or other regularly shaped volumes) are the periodic (or toroidal) boundary conditions. In a d-dimensional box with linear dimensions $L_1, L_2, ..., L_d$ the boundary conditions are specified by requiring that the variables $s(\boldsymbol{x})$ are periodic:

$$s(\boldsymbol{x}) = s(\boldsymbol{x} \pm \boldsymbol{L}_\nu) \quad , \qquad \text{for all} \qquad \nu = 1, ..., d \quad . \tag{2.43}$$

Thus the sites on the uppermost and lowermost boundaries of a two-dimensional rectangle are considered to be neighbours, as are the sites on the left- and right-hand edges (or the uppermost and lowermost planes of a three-dimensional lattice, and so on). This is particularly convenient for problems involving local interactions, while for systems that involve long-range potentials (such as the Coulomb $1/r$ potential) a cutoff may need to be introduced.

Other boundary conditions are sometimes used in particular circumstances. The simplest of these are free boundary conditions, in which the edges of the box or lattice represent the physical extent of the system and the "missing" sites are given a value of zero. This is useful if, for example, we want to study free surfaces of a system. Antiperiodic boundary conditions are defined simply by putting a minus sign on the right-hand side of (2.43), and may be used for example to simulate the coexistence of two phases with opposite values of the order parameter.

The effect of the boundary conditions is most evident when the correlation length ξ becomes large. The singularities that one expects to see on an infinite system in the neighbourhood of a continuous phase transition are rounded off. These and other finite-size effects have received considerable attention and will be discussed further when we look at finite-size scaling in Sect. 2.5.4.

2.5.2 Effects of Finite Simulation Time

In the expression for the measurement of a thermodynamic quantity given in (2.9) we have assumed that the configurations being measured are in equilibrium. In practice it is often convenient to start with either a totally ordered configuration or one in which the spins take on random orientations, corresponding to low and high temperatures respectively. The initial configurations generated will in general not represent the equilibrium states at the temperature we wish to simulate, and therefore they must be excluded from the average (2.9). It may sometimes be difficult to distinguish slow relaxation or metastable behaviour from true thermal equilibrium, a feature that is, for example, characteristic of spin glasses. One method for deciding if a system is in equilibrium is to compare measurements from runs which have different starting configurations (e.g., hot and cold starts). When the results agree within the statistical error then we can be confident that the effect of the starting configuration has been eliminated and that the subsequently generated configurations represent thermal equilibrium.

In the neighbourhood of a first-order phase transition metastable behaviour may be observed. It is often the case that it is not possible to run the simulation long enough to see the metastable state decay into true equilibrium. Such behaviour can nevertheless be recognised by comparing simulations starting from hot and cold initial configurations. The true equilibrium can be identified as those states with the lower free energy.

The next question that we need to consider is the length of simulation time, i.e., the number of configurations necessary in the average of (2.9) in order to achieve the required accuracy. In Sect. 2.2.3 we saw that the Monte Carlo process can be viewed dynamically and that it is possible to measure the correlation τ between subsequent configurations. The naive error estimate for a measurement of a quantity A is modified by the correlations between successive configurations according to (2.21). In the neighbourhood of a continuous phase transition this autocorrelation time τ_A will diverge with the *dynamical critical exponent* z, conventionally described by [2.73,74]

$$\tau \propto \xi^z \propto |1 - T/T_c|^{-\nu z} \quad , \tag{2.44}$$

where ξ is the correlation length. This phenomenon is known as *critical slowing down* and plays a central role in any discussion of simulation algorithms. For simulations carried out at a critical point we expect this divergence to be rounded off on a system of finite volume and give, in general, a finite-size dependence of the autocorrelation time of the form

$$\tau \propto L^z \quad . \tag{2.45}$$

The exponent z is notoriously difficult to measure and it requires considerable expenditure of computer resources. Among the models for which z has been measured to date are the Ising model with Glauber and/or Metropolis dynamics [2.75–77], the Ising model with conserved energy and Q2R dynamics [2.15] and the Ising model with conserved energy and Creutz dynamics [2.78]. A quite different behaviour is, however, seen when cluster-update methods are used to study the two-dimensional Ising model, as is discussed in Sect. 2.6.3.

2.5.3 Statistical Errors and Self-Averaging

In any simulation of a physical system the measurement $\overline{A(s)}$ (2.9) will deviate from the expectation value $\langle A(s) \rangle$ (2.3) due to both the finite size of the system and the finite number of configurations. The statistical error on a system of linear dimension L in which we have n *independent* measurements is given by

$$\Delta(n, L) = \sqrt{(\langle A^2 \rangle_L - \langle A \rangle_L^2)/n} \quad , \qquad n \gg 1 \quad . \tag{2.46}$$

As we saw in the previous section however, when a sequence of configurations is generated using the Monte Carlo procedure then the successive measurements we make will not, in general, be independent of each other. It is possible to measure the correlation time explicitly using (2.17) for the normalized correlation function, but a method more frequently used in practice is *binning*, which proceeds as follows. Choose an initial trial estimate for the correlation time – say m consecutive configurations. Now form $n_I = M/m \gg 1$ averages of the measurements over the m consecutive configurations. We may then check whether the error on these n_I averages decreases as $1/\sqrt{n_I}$. If this is not the case then the correlation time must be chosen to be larger, i.e., a larger value of m must be chosen and the above steps repeated.

Consider now the *finite-size behaviour of statistical errors*. A useful concept is that of *self-averaging* [2.2,30]. If the error $\Delta(n, L)$ of the quantity A vanishes as L becomes larger and larger then we say that A is self-averaging, whereas if Δ has a non-zero limit A exhibits a lack of self-averaging. We may also ask about the relationship between the relative errors obtained on volumes of different sizes – and particularly whether in order to reduce the error it would be better to invest a certain amount of computer time in studying either larger systems or a larger number of smaller configurations. We can now compare calculations on two systems with linear dimensions L and $L' = bL$, where $b > 1$ is a scale factor. Suppose we do n measurements on the smaller system and $n' = b^{-d}n$ measurements on the larger system (d is the dimension of our space). The system is called *strongly self-averaging* if the error satisfies the relation

$$\Delta(n, L) = \Delta(n', L') = \Delta(b^{-d}n, bL) \ . \tag{2.47}$$

It then follows from (2.46) that strong self-averaging holds only if

$$\langle A^2 \rangle_L - \langle A \rangle_L^2 \ \propto \ L^{-d} \ . \tag{2.48}$$

In general this is true for measurements of quantities such as the energy per site (E) and magnetization per spin (M) that correspond to the density of a basic extensive quantity, as long as $L \gg \xi$ (where ξ is the correlation length).

However, the situation is very different for measurements of quantities like the specific heat (C) and magnetic susceptibility (χ) that follow from fluctuation relations. In such cases it turns out that increasing L at fixed n produces *no* gain in the accuracy at a critical point of such quantities as C and χ, although it will greatly improve the accuracy of quantities like E and M. It may therefore be better to obtain C, χ, etc., from numerical differentiation rather than from the sampling of fluctuations. For a more comprehensive discussion of this point the reader is referred to [2.2,30].

2.5.4 Finite-Size Scaling: Using Finite-Size Effects

A system of finite size will give a good approximation to an infinite system as long as the correlation length ξ is smaller than the linear dimension L of the system. However, when we look at systems for which $\xi \geq L$, as is the case near a continuous phase transition, the results will in general be strongly dependent upon the nature of the boundary conditions. The singularities that would be expected on an infinite system, such as the specific heat and magnetic susceptibility, are then both *rounded* and their peaks slightly *shifted* in temperature.

The central idea of the finite-size scaling hypothesis [2.79,80] (see also [2.72] for a review) is that the behaviour of a system with linear dimension L near the infinite volume critical temperature T_c is determined by the scaling variable

$$y = L/\xi \quad , \tag{2.49}$$

where ξ is the temperature-dependent correlation length of the infinite system. We may define an "effective critical temperature" $T_c(L)$ on a finite system in some reasonable way such as, for example, where the peak in the specific heat lies. Finite-size scaling theory then tells us that the shift in critical temperature scales as

$$\delta T_c = T_c(L = \infty) - T_c(L) \sim L^{-1/\nu} \quad , \tag{2.50}$$

where ν is the usual critical exponent describing the divergence of the correlation length near T_c. The magnetization (or, in general, the order parameter scaling function) has the functional form

$$M \sim L^{-\beta/\nu} f(hL^{\beta\delta/\nu}, tL^{1/\nu}) \quad , \tag{2.51}$$

where h is an external magnetic field and t is the reduced temperature $t = (T - T_c)/T_c$. Likewise the susceptibility behaves as

$$\chi \sim L^{-\gamma/\nu} g(hL^{(\gamma+\beta)/\nu}, tL^{1/\nu}) \quad . \tag{2.52}$$

Thus a plot of $ML^{\beta/\nu}$ vs $tL^{1/\nu}$ for $h = 0$ and various values of t and L gives a scaling function, as does a plot of $ML^{\beta/\nu}$ vs $hL^{\beta\delta/\nu}$ for $T = T_c$ and various values of h and L. This method enables us to determine the critical exponents ν, β, etc., although the actual form of the scaling functions are not universal and depend upon the choice of boundary conditions. Finite-size scaling theory has been successfully used in a wide variety of problems in statistical mechanics and it has been found that such simulations frequently require only relatively small volumes, which are well within the grasp of even moderately sized computers today.

2.6 Monte Carlo Algorithm for the Ising Model

In order to illustrate the algorithms we discuss, we have chosen one of the simplest and most widely used models in statistical mechanics – namely the Ising model. In this section we describe the model and give an outline of how a Monte Carlo update program for measuring simple equilibrium properties of the Ising model is structured. A complete program, written in Occam, is found in Appendix A. The equivalence of the Ising model to bond percolation, as first derived by *Kasteleyn* and *Fortuin* [2.81,82], is outlined. This equivalence enables us to *rewrite* the Ising model using an algorithm due to *Swendsen* and *Wang* [2.23] based on *clusters*, which considerably lessens the correlation between successive Monte Carlo configurations, thus reducing the effect of *critical slowing down*. This algorithm, although it is computationally more difficult, consequently warrants closer attention, and an outline of such a cluster update program for the Ising model is given.

2.6.1 The Ising Model

The Ising model was originally introduced by *Lenz* [2.83] in 1920 and the one-dimensional model, which has no finite temperature phase transition, was solved by *Ising* [2.7] in 1925. *Onsager* [2.84] succeeded in 1944 in solving the two-dimensional Ising model in zero magnetic field, which he showed has critical behaviour, while the three-dimensional Ising model has until now yielded only to approximate solutions, series expansions, real-space renormalization group calculations and computer simulations; it remains a model of great research interest.

The Hamiltonian for the Ising model in an external magnetic field H is

$$\mathcal{H}_{\text{Ising}}(s) = -J \sum_{\langle ij \rangle} s_i s_j - H \sum_i s_i \quad , \qquad s = \pm 1 \quad , \tag{2.53}$$

where $\langle ij \rangle$ are all nearest-neighbour pairs of lattice sites, H is the external magnetic field, and the exchange coupling J is restricted to be positive. The spins take only the values of +1 (spin-up) or −1 (spin-down). It is remarkable that such a simple model displays such rich behaviour and that it is capable of describing a wide variety of physical systems, ranging from alloys, magnetic materials, polymers and fluids to biological membranes, and so on. The fact that such a simple model can describe such diverse behaviour is bound up with the concept of *universality*, in which the large-scale critical behaviour of the model depends less upon the detailed microscopic interactions than upon more basic characteristics of the interactions, such as their symmetry. In our examples using the Ising model we will draw upon several of the above physical systems for illustration and motivation – e.g., we can think of the Ising model as describing a ferromagnet with a very strong uniaxial anisotropy with spins pointing up and down along the "easy axis", or as a percolation problem with spin-up and spin-down corresponding respectively to occupied and unoccupied sites, or a binary alloy with spins corresponding to the two sorts of atoms, and so on.

2.6.2 Implementing the Monte Carlo Algorithm for the Ising Model

We outline here how the Monte Carlo algorithm is implemented for the two-dimensional Ising model on an $L \times L$ lattice with periodic boundary conditions. The structure of the algorithm given here is applicable to any sort of computer architecture – scalar, vector or parallel. A specific implementation of this program in Occam is given both as a serial program (Sect. 9.2.5) and including parallel features [2.85] (Appendix A.).

In doing a Monte Carlo update it is possible to move through the lattice sites updating the spins in either a sequential or a random fashion. For equilibrium properties both methods are equally good – we choose to visit the sites in a regular (typewriter) fashion because it is considerably faster. However, if we want to measure dynamical properties of the system a random sequence of visiting sites is more appropriate. In such a case it is important that a good pseudorandom number generator is used, otherwise it may happen that certain sites are never visited and the results of the simulation are consequently meaningless. Combinations of regular and random updating patterns are also perfectly feasible.

We start by taking a spin configuration, which has either come from a previous simulation or which we initialize explicitly. A cold initial configuration is one in which all spins are in the same direction, and a hot configuration is one where the spins are chosen to be ± 1 at random.

Algorithm: The Monte Carlo update of an Ising spin

One lattice site i with spin s_i is considered for updating.

- Calculate the change in the Hamiltonian $\Delta\mathcal{H}$ between the existing configuration $\mathcal{H}^{\text{old}}$ and the configuration $\mathcal{H}^{\text{trial}}$ with this single spin flipped $s_i \rightarrow -s_i$: $\Delta\mathcal{H} = \mathcal{H}^{\text{trial}} - \mathcal{H}^{\text{old}}$.
  Either:

  1. $\Delta\mathcal{H} \leq 0$. The spin variable s_i is replaced by its flipped value $-s_i$.
  Or:

  2. $\Delta\mathcal{H} > 0$. Then calculate the transition probability for flipping the spin: $\delta = \exp(-\Delta\mathcal{H}/k_{\text{B}}T)$. Draw a (pseudo)random number R, which has a uniform distribution between 0 and 1.
  Then

     - $\delta < R$ leave spin s_i unchanged, or
     - $\delta \geq R$ flip the spin s_i to become $-s_i$.

The spin at site i is now *updated* (although, of course, it may not have changed from its original value) and we move to the next site and repeat the above steps. One *lattice-update* (or *sweep*) corresponds to one *spin-update* of every spin on the lattice (i.e., every spin site is visited once).

Note that in the case $\Delta\mathcal{H} \leq 0$ the spin is automatically flipped and it is not necessary to generate a random number. A further simplification occurs by using the fact that only a very small number of possibilities for $\Delta\mathcal{H}$ exist – in our example here there are only a total of five possibilities for the Ising model in two dimensions. Thus it is not necessary to calculate δ each time, since it suffices to have a *look-up table* where the five possible values are stored. This has the advantage that the exponential does *not* need to be calculated for every spin-update, which saves considerable computer time. Such a look-up scheme works for all problems where the number of possible values of δ is finite, as is the case for systems with discrete degrees of freedom.

Thermodynamic quantities, such as the magnetization, can then be measured after each lattice update. In the case of the magnetization it is particularly simple to keep a running value M_{tot} that, each time a spin s_i is flipped, is replaced by $M_{tot} + 2s_i$. As was discussed earlier in Sect. 2.5.2, it may be more efficient not to carry out measurements after each lattice update, but perhaps only after the successively generated configurations are sufficiently independent. It may also be necessary to measure their correlations explicitly, as described in Sect. 2.5.2.

2.6.3 The Swendsen-Wang Algorithm and the Equivalence Between the Ising Model and Percolation

In this section we outline a deep correspondence between the Ising model (or, more generally, the q-state Potts model) and bond percolation, which was originally described by *Kasteleyn* and *Fortuin* [2.81] in 1969. But before we discuss this connection it is first necessary to say a few words about percolation. The most straightforward type of site percolation consists of a lattice in which each site can be either occupied or vacant. A site is occupied with a probability $p \in [0, 1]$ and vacant with probability $1 - p$. On an infinite lattice (i.e., in the thermodynamic limit) there is a certain probability p_c above which an infinite *cluster* can exist. A cluster is defined by the condition that two nearest-neighbour sites that are occupied belong to the same cluster. Analytic results for this percolation threshold p_c are known only in two (and infinite) dimensions, and thus computer simulations play an important role in understanding these models. We will be interested here in a slight variation of this simple site percolation model, called bond percolation, in which it is the bonds that are occupied or vacant and the clusters consist of connected bonds. A review of percolation has been given by *Stauffer* [2.86].

Our interest in this correspondence between bond percolation and the Ising model is because it enables us to reformulate the Ising model in terms of a cluster algorithm. This algorithm, due to *Swendsen* and *Wang* [2.23] has the desirable property that its dynamical critical exponent z is unusually small. This means that, on large lattices, it is possible to generate independent equilibrium configurations much more efficiently. One crucial feature of this algorithm is that it deals with *clusters*, and is therefore *non-local* in character.

The idea that *clusters* play an important role in spin systems has a long tradition. Clusters have a central role, for example, in *Fisher*'s droplet model [2.87], which provides a phenomenological description of continuous phase transitions. It is useful here to introduce the idea of *geometric clusters* (as opposed to the *physical clusters* that we shall consider shortly), in which all neighbouring spins that are parallel belong to the same cluster. However, when we use this geometric definition of clusters to investigate the critical properties of the Ising model, we find that in three dimensions they *percolate* (i.e., the largest cluster spans the system) at a temperature *below* the Ising critical temperature. It was not until considerably later that *Coniglio* and *Klein* (1980) [2.88] and *Hu* (1984) [2.89] introduced the idea of a temperature-dependent bond occupation factor $p = 1 - \exp(-2J/k_B T)$, by which it is possible to define *physical clusters* as being parallel spins connected by occupied bonds. (This idea is implicit in the earlier work of *Kasteleyn* and *Fortuin* [2.81] showing the analytic correspondence.)

In the outline that we give here of the identity between the Ising model and bond percolation we keep as close as possible to the spirit of the Swendsen-Wang algorithm. We also introduce explicitly an external magnetic field and show how the partition function for the bond percolation model is thus modified.

We begin by considering an Ising spin configuration s, which occurs with a probability given by its Boltzmann factor (2.1) with the Ising Hamiltonian (2.53). We build from it a bond configuration Γ (the bonds being simply the links between nearest-neighbour sites) according to the following rules:

- All bonds between anti-parallel spins are *open*.

- Bonds between parallel spins are either

  - *open* with probability $q = \exp(-2J/k_B T)$
    or
  - *closed* with probability $p = 1 - q = 1 - \exp(-2J/k_B T)$.

Thus, the bond configuration Γ is a set of open and closed bonds, which are donated by $o(\Gamma)$ and $c(\Gamma)$ respectively. The reason for the given value of the bond probability p will become apparent shortly. The probability of the bond configuration Γ is then

$$P(\Gamma) = \sum_s P(s)\, P(\Gamma|s) \quad , \tag{2.54}$$

where

$$P(\Gamma|s) = \delta_{\Gamma,s}\, p^{c(\Gamma)}\, q^{N_p(s)-c(\Gamma)} \tag{2.55}$$

is the conditional probability that the bond configuration Γ is generated from the spin configuration s, and $N_p(s)$ is the set of all bonds that lie between two parallel spins. The function $\delta_{\Gamma,s}$ is unity when the configurations s and Γ are compatible (in the sense of the above rules) and is zero otherwise.

We now use this bond configuration Γ to build a new spin configuration s' according to the following rules:

- First use the bonds to build *clusters* – all spins connected by closed bonds belong to the same cluster.

- Each *cluster* is now given a new orientation according to the following:

  - A cluster takes spin -1 with probability r

$$r = \frac{1}{1 + \exp[2HN(\lambda)/k_B T]} \quad .$$

  - A cluster takes spin $+1$ with probability t

$$t = 1 - r = \frac{\exp[2HN(\lambda)/k_B T]}{1 + \exp[2HN(\lambda)/k_B T]} \quad ,$$

  where $N(\lambda)$ is the number of spins contained in the cluster λ and all the spins within the cluster take this new orientation.

The probability of the new spin configuration s is likewise given by

$$P(s') = \sum_{\Gamma} P(\Gamma) \, P(s'|\Gamma) \quad , \tag{2.56}$$

where $P(s'|\Gamma)$ is the conditional probability that the spin configuration s' is generated from the bond configuration Γ

$$P(s'|\Gamma) = \delta_{\Gamma,s'} \, r^{\gamma^-(\Gamma)} \, t^{\gamma^+(\Gamma)} \quad . \tag{2.57}$$

As before, the function $\delta_{\Gamma,s'}$ is unity when the configurations s' and Γ are compatible (in the sense of the above rules) and is zero otherwise. $\gamma^-(\Gamma)$ and $\gamma^+(\Gamma)$ are the number of clusters in the configuration Γ with spin -1 and $+1$ respectively. It is straightforward to check that both $P(\Gamma|s)$ and $P(s'|\Gamma)$ are stochastic matrices:

$$\sum_{\Gamma} P(\Gamma|s) = \sum_{s'} P(s'|\Gamma) = 1 \quad . \tag{2.58}$$

The partition function for bond percolation can now be written analytically as

$$\tilde{Z} = \sum_{\Gamma} p^{c(\Gamma)} \, q^{N_B - c(\Gamma)} \prod_{\lambda}^{\gamma(\Gamma)} (1 + e^{2HN(\lambda)/k_B T}) \quad , \tag{2.59}$$

where the product is over all clusters λ contained in the configuration Γ. The tilde over the Z indicates that we have shifted the Hamiltonian by a trivial constant, equal to $JN_B - HN_S$, where N_B is the total number of possible bonds on the lattice and N_S is the number of sites. This is the result of *Kasteleyn* and *Fortuin* [2.81], modified to include the effect of the external magnetic field. *Sweeny* [2.90] has used this expression (but without the magnetic field term) to carry out a Monte Carlo simulation of the Ising model by generating percolation configurations with the appropriate weight.

The value of $q = \exp(-2J/k_B T)$ has been so chosen that, given that $P(s)$ has a Boltzmann probability distribution, the configuration s' also has the canonical probability distribution:

$$P(s') = \sum_{\Gamma} \sum_{s} P(s'|\Gamma) \, P(\Gamma|s) \, P(s) = \frac{1}{Z} \exp[-\mathcal{H}(s')/k_B T] \quad . \qquad (2.60)$$

Moreover, the spin clusters that are defined by the bonds percolate at the Ising critical temperature and, as we see from the construction of the new spin configuration s', the clusters do *not* interact with each other. For these reasons they are referred to as *physical clusters*, in contrast to the *geometric clusters* introduced earlier. The effect of introducing the bonds has been to split up some of the geometric clusters into smaller clusters, and this effect becomes more pronounced at higher temperatures, where the probability of closed bonds becomes correspondingly smaller.

Thus it is possible to generate successive Ising spin configurations s and s' using the bond configuration Γ as an intermediatory step. Equation (2.60) ensures that the Boltzmann probability distribution is an eigenvector of this two-step process, and thus that the Markov chain of spin configurations converges to the canonical probability distribution. *Swendsen* and *Wang* [2.23] originally proposed this algorithm and measured a dynamical critical exponent of approximately 0.35, which is roughly an order of magnitude less than that for the conventional Metropolis algorithm. As we saw in the discussion of finite-size effects and critical exponents in Sect. 2.5.2, this provides an enormous reduction in the correlation time between configurations, especially for simulations on large lattices near the critical point. Consequently it is possible to measure quantities with far greater precision and to carry out more accurate simulations than is possible with the local Metropolis update algorithm. The price that is paid for this gain is that each lattice update now involves the identification of all the clusters generated by the intermediate bond configuration, and this is a very non-local procedure which involves the possible connection of spins that are spatially widely separated. We give here an outline of one Monte Carlo lattice update using this algorithm, and in the following section discuss how clusters can be efficiently handled in such a simulation.

Algorithm: One lattice update using the Swendsen-Wang algorithm

---

We start with a configuration of Ising spins $s = \{s_i\}$.
- Construct a bond configuration Γ as follows:

  - Bonds between anti-parallel spins are open.
  - Bonds between parallel spins are open with probability $q = e^{-2J/k_B T}$, and closed with probability $p = 1 - q$.
- Identify all the clusters λ_i in this bond configuration (details of methods to do this follow below).
- Each cluster λ_i is given a new spin configuration of -1 with probability $r = 1/(1 + e^{2HN(\lambda_i)/k_B T})$ or $+1$ with probability $t = 1 - r$. All the spins in the cluster λ_i take this new spin value [$N(\lambda_i)$ is the number of spins in cluster λ_i].

This gives a new spin configuration s', on which the above sequence of steps can be repeated.

The clusters that are generated at the intermediate step of this algorithm can contain anything from a single spin to a set of spins that could (contiguously) reach from one side of the lattice to the other. The fact that these clusters can be so spread out means that each lattice update takes considerably longer than for a local update algorithm such as the standard Metropolis method outlined earlier. This disadvantage is, at least for large lattices, outweighed by the reduction in the effect of critical slowing down for temperatures in the neighbourhood of the critical point [2.91].

2.6.4 Cluster Identification

The question of how to identify clusters deserves special attention. The method we describe, which is suitable for any scalar machine, is due to *Hoshen* and *Kopelman* [2.92] and was further developed by *Kertesz* [2.93]. Basically the algorithm is a labelling technique, and we describe it in the context of the Swendsen-Wang algorithm for the Ising model with intermediate bond configurations that define the clusters – although the method is more generally useful for all types of cluster identification problems. A fuller discussion of cluster identification algorithms is given by *Stauffer* [2.86] (Appendix A.3).

The algorithm starts out with the site in the "upper left corner" of the lattice. For simplicity we also assume that the boundaries are free, i.e., that sites on the boundary cannot be connected to the outside. Since this is the first site we give it the cluster label number 1. We also define a "stack vector" (or "cluster permutation vector") whose first element corresponds to this first cluster and is assigned the value -1 or -3 (for spin values $+1$ and -1 respectively). Each spin on the lattice is sequentially (in typewriter fashion) visited. If it is not connected with any labelled cluster it is assigned a new label and the corresponding spin-value entry placed in the stack vector. Any spin that is connected to only one labelled cluster is assigned that particular cluster label. Suppose now that a spin is connected to two labelled clusters. In this case a joining of the two clusters occurs, as shown in Fig. 2.1. The spin is assigned the lowest of the two cluster labels, and the other cluster with which this spin has now connected is joined by overwriting its (negative) spin value in the stack vector with this (positive) lowest cluster label (on a single processor this condition of "lowest cluster number wins" is not necessary, and is only for book-keeping purposes). When the next cluster is joined in this manner the algorithm iterates through the stack vector, which serves as a linked list, until it reaches the "seed cluster", which is the last cluster in this linked chain (and thus the one with the lowest cluster label). After one sweep through the lattice all clusters are stored in the stack vector.

If we are only interested in labelling the clusters it is possible to do so without introducing the stack vector in the above algorithm. We have introduced the stack vector here because it proves to be useful later when we discuss how such cluster identification algorithms can be carried out on a lattice that is distributed over a number of computing processor elements. However, in the straightforward case of cluster identification on a single scalar processor the algorithm can be simplified. In this case it is necessary to have some implicit numbering of the lattice sites. Again we move through the lattice in typewriter fashion giving each new cluster a *negative* cluster number and each site that is connected to an existing cluster receives the site number of the seed of the cluster. When a site connects two clusters, one of the cluster seeds must be overwritten by the site number of the other seed, thus generating a linked list. Indeed, in some contexts cluster numbers may be superfluous and the seed of the cluster can retain the spin orientation of the cluster.

Clearly the problem of the identification of the clusters is non-local. This cluster labelling procedure enables two clusters which were originally widely separated to be part of a single larger cluster. The problem of cluster identification on vector machines is, however, rather problematic. The method that is outlined here is not suitable for vectorization since the permutation vector, which essentially stores connections between the branches of the various clusters, may be updated in a non-trivial way by each of the spins as we move through the lattice. This means that at every lattice site there is an unpredictable number of backward jumps in order to reach the seed of the tree of assignment labels. Nor have any other methods more amenable to vectorization been found that handle the cluster identification in an efficient way. The difficulty seems to lie in the nature of the problem of cluster identification, which is both non-local and inhomogeneous. This method does, however, lend itself to parallelization, which is discussed in some detail in Sect. 7.3.

2.6.5 Other Cluster Update Algorithms

The success of the Swendsen-Wang algorithm in reducing the effect of critical slowing down in the neighbourhood of a critical point has led to the search for

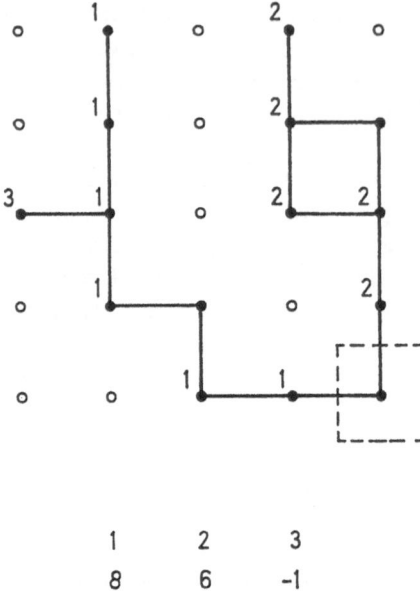

1	2	3
8	6	-1

Fig. 2.1. A sample configuration of spin-up and spin-down sites where two large clusters merge. The numbers show the labels assigned by the algorithm

cluster algorithms that are applicable to models with continuous fields. In this section we will discuss the method of *Wolff* [2.24], as modified from a one-cluster to a many-cluster algorithm [2.94], which is more appropriate for a simulation on parallel processors. The one-cluster algorithm has been used in a number of high statistics investigations of the X-Y model [2.95], the O(3) sigma model [2.96] and the O(4) model [2.97]. A number of other earlier generalizations of the Swendsen-Wang algorithm for continuous field models [2.98–101] suffer from various drawbacks, whereby the whole lattice tends to end up in one cluster, or the surface energy of the clusters is so large that the clusters rarely undergo any change, or the algorithm simulates a variant action whose relationship to the original model is uncontrolled. We also mention in passing that there are a number of other methods that are being developed to overcome the problem of critical slowing down. Multigrid methods have been successful in this regard [2.20–22], as have Fourier acceleration techniques [2.19] and overrelaxation algorithms [2.17,18], and the reader is referred to the articles cited for more details concerning these developments.

We consider the O(n) model, which is defined by the Hamiltonian

$$\mathcal{H} = -\beta \sum_{\langle ij \rangle} \boldsymbol{s}_i \cdot \boldsymbol{s}_j \quad , \tag{2.61}$$

where $\boldsymbol{s}_i$ is a unit vector in $\mathcal{R}^n$ and the sum is over all nearest neighbours.

Algorithm: Cluster update algorithm for the $O(n)$ Model

We start with a configuration of continuous spins $\{s_i\}$.

- Construct a bond configuration Γ as follows:

  - The bond $\langle ij \rangle$ is deleted with probability
    $q = \exp[-\beta(s_i^1 \cdot s_j^1 + |s_i^1 \cdot s_j^1|)]$.
  - The bond $\langle ij \rangle$ is frozen with probability $p = 1 - q$.

- Identify all the clusters λ_i in this bond configuration (exactly as previously for the Ising model).

- In each cluster λ_i all the spin components s^1 change sign with probability $1/2$.

- Do a global transformation of all the spins on the lattice in such a way that the product of a fixed number of such rotations gives a completely arbitrary rotation in the spin-space.

This sequence of steps gives one update of the spin variables.

This algorithm satisfies the requirement of detailed balance and ergodicity. The one-cluster algorithm differs in that initially only one site on the lattice is chosen randomly and then only the cluster associated with this site is constructed. All the spins in this cluster are then reflected, i.e., their one-component is flipped. A comparison of the dynamical behaviour of these two algorithms [2.36] shows that they do not coincide even for the case of the Ising model.

A further advantage that cluster algorithms offer is that it is possible to construct improved estimators [2.90], which give a reduction of the variance of physical quantities. This follows from the observation that a cluster configuration corresponds to a number of different spin configurations, which can therefore be straightforwardly averaged when taking measurements.

3. Physics and Parallelism

Many of the methods employed to solve problems, both in physics and in other branches of science, lend themselves naturally to parallelization. This statement, although it may seem obvious to many readers, needs some clarification. To what extent is this statement true and are there problems which are inherently serial? We may indeed suspect that there are some problems which are not susceptible to parallelization. We have already encountered the exponential function, which we identified as being serial. It is also possible that some problems are amenable to parallelization but that it is not optimal to solve them in such a way.

We start our investigation by examining some real-life problems and discuss their apparent or inherent parallelism. The two most prominently featured models in this text are the Ising model

$$\mathcal{H}_{\text{Ising}}(s) = -J \sum_{\langle i,j \rangle} s_i s_j - H \sum_i s_i \tag{3.1}$$

and the N-particle system with the Hamilton function

$$\mathcal{H} = \frac{1}{2} \sum_i m v_i^2 + \sum_{i<j} V(r_{ij}) \quad , \tag{3.2}$$

which were described in detail in the previous chapter (Sects. 2.3 and 2.6).

Many-particle systems always involve interaction and the question is whether the interaction can be exploited for parallelization. A system of particles inside a box is shown in Fig. 3.1. The particles interact with each other by potentials which we assume to be short ranged, in the sense that they have a definite cutoff well below the linear extension of the box. Each particle can interact only with a subset of the other particles in the system. This observation allows us to partition the system into cells such that only the nearest and next-nearest neighbour cells can interact. Of course such a division is only one possible partitioning of the particles into quasi-independent segments. The point which we want to stress is that it is almost always possible to break up a physical system into geometric segments which do not directly interact with each other.

In general the system will be time-dependent and the configuration will continuously evolve as time proceeds. For this reason the partitioning cannot be too rigid. Let us assume for the moment that we look at such a system at a fixed time. We are, of course, forced to consider the communication between the cells into which the system under consideration has been partitioned. Each cell can be associated with a processor, but due to the nature of the problem there will of course be a need for interprocessor message exchange to ensure the correct physical behaviour. Consequently there must be appropriate message paths between the processors. In the above partitioning in two dimensions each cell has eight neighbouring cells and with each of these eight neighbouring cells there may be interactions, which requires a message exchange.

We can parallelize the Ising model along similar lines. The short-range nature of the Ising interaction can be exploited to partition the underlying lattice. Several such partitionings are possible. One way to cut the lattice into quasi-independent parts is to use sublattices. The simplest such partitioning of the lattice is to split it into two interpenetrating sublattices [3.1,2]. Each of the sublattices has a lattice spacing such that neighbouring spins do not interact directly with each other and hence can be worked upon concurrently.

In general we can say that interaction makes possible parallelization in a geometric sense. The real physical space on which the model is defined is decomposed into parts according to the interaction among the entities. There must however be some communication, since the parts are not completely independent of each other. *Causality forces communication between the units on which a parallel action is performed.* In the example of the N-particle system it was the exchange of the particle positions and velocities of the nearest and next-nearest neighbour segments. Causality does not require any direct communication with parts of the system further away. Indirectly some communication is necessary, since the configurations change in time and particles leave their neighbourhood.

In the example of the Ising model the communication necessary to ensure a correct causality is more complicated. Suffice it to say here that we must fulfil detailed balance to ensure a proper distribution and generation of the Ising configurations. As will be elaborated in later chapters, this will impose additional constraints on the way we partition the lattice and which parts need to communicate. On the other hand, the fact that the spins do not move in space but remain fixed on their lattice sites means that there are considerable simplifications over the N-body problem.

Here we see that one of the limiting factors for the speed-up which we can attain will be the requirement of causality. Only for a negligible amount of communication compared with the amount of computation can we expect to enter a regime of asymptotic linear speed-up. With the requirement of causality we have a built-in limiting factor on what can be achieved with respect to a reduction in the computational complexity.

It is not a necessary condition for parallelism that a system has short-range interactions. Even a system where every particle interacts with every other particle in the system is capable of being parallelized. In such a case the burden is shifted almost entirely onto the communication between the processors. In

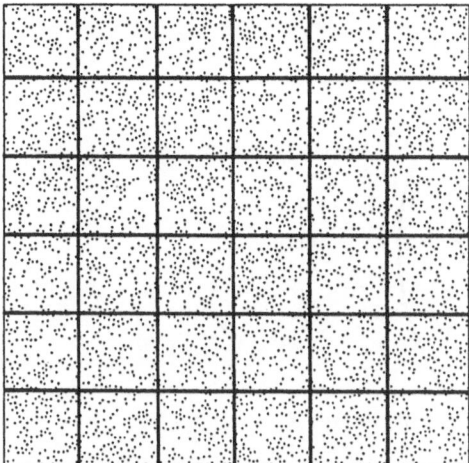

Fig. 3.1. A many-particle system inside a box. The box is divided into sections such that each segment interacts only with its nearest and next-nearest neighbours

Chap. 8, which concerns data parallel algorithms, and in Appendix C we discuss a program for such a situation in some detail.

It is a straightforward but by no means trivial observation that for many of the problems we encounter in computational science we need to investigate a set of independent parameters $p_1, ..., p_m$. This set might consist of temperatures to scan a phase diagram, a set of different system sizes, a set of interactions and so on. All these parameters can be investigated simultaneously since they are independent. We can write this in a formal way as

$$\text{some } p \quad : \text{some } x_0 \quad : \mathcal{A}(p, x_0) \quad . \tag{3.3}$$

This idea can be elaborated further. Let us consider the percolation problem [3.3]. The partition function for the percolation problem is given by

$$Z = \sum_s B(s, p)2^{c(s)} \quad . \tag{3.4}$$

Each configuration $s \in \Omega$ consists of sites of a lattice which are either occupied, with a probability p, or empty. Each site is occupied independently of the other sites. It is thus possible to generate a configuration of occupied/empty sites in parallel. A more serious problem is, of course, to analyse the configuration. The parallel analysis of such cluster configurations will occupy us later in Sect. 7.3.

A problem where the constituents are at least for some period of time independent is the diffusion limited aggregation model [3.4]. In Fig. 3.2 the basic process is shown. Independent random walkers are introduced into the system and begin their walks. Initially there is one seed-site and any walker which arrives at this site (i.e., reaches a nearest neighbour site of the seed) remains there and acts as an additional seed-site. The important point which we want to emphasize in this example again concerns causality. The walkers are independent

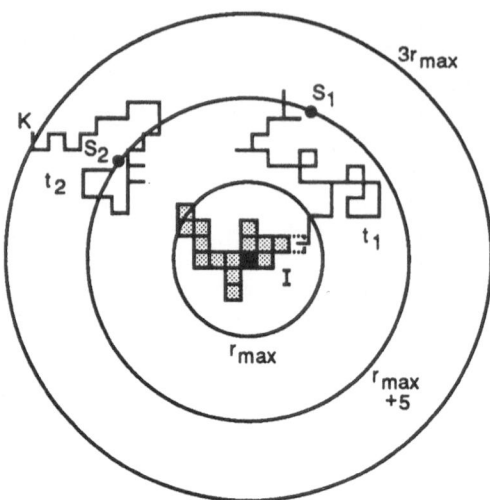

Fig. 3.2. Diffusion limited aggregation process

when they come into the system and they remain so as long as they do not become neighbours. As soon as two walkers become neighbours they influence each other's paths. It is only then that one needs to introduce a mechanism for synchronization. Otherwise they are independent entities which can be handled in parallel.

As a further example of this idea consider a polymer problem [3.5–8], and in particular consider the generation of a new conformation of a single polymer chain from a previous conformation. A conformation of a chain on a lattice is shown in Fig. 3.3. Ideally we would like to move all beads of the chain to a new position, but the geometrical constraint which requires the non-overlap of beads clearly prohibits a simple independent move of all beads. However, we could, as before, partition the underlying space into cells which are independent. Beads lying in cells which cannot possibly influence each other can be moved in parallel.

While a physics problem itself will often be inherently parallel, we must understand the parallelism (or its absence) resulting from either the formulation of the problem or the method of solution. Two formulations of the same problem may result in a vastly different complexity with respect to the amount of computation necessary and type of parallelization. It is to this aspect of parallelization in computational science problems that we want to give more consideration in the next chapter. One point, however, can be made right away.

The approach taken so far, as in the example of the particle in the box problem, was that we considered the interaction as the lever for the parallelization. We used this lever to break up the system into quasi-independent parts with time being the driving force by which the system was propelled. Interaction can be used to parallelize the system in a different way. As long as two particles are

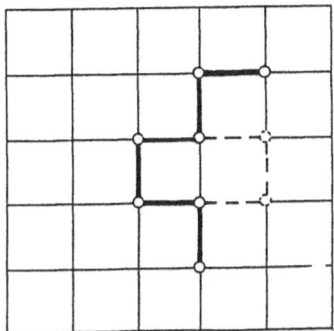

Fig. 3.3. Conformation of a single polymer chain on a lattice. The change to a new conformation as proposed, for example, by a Monte Carlo method, is indicated

further apart than the cutoff, they are independent and move freely. To illustrate this situation more dramatically we consider the hard-sphere problem [3.9,10]. Only when two particles collide do they change their paths. Instead of driving the system by time it is possible to consider the system as being driven by the collision events [3.11]. The system will then not be parallelized geometrically but rather according to the events which take place.

In general we can say that there are several types of parallelism inherent in physics problems. These are:

- independence

- time correlation

- space correlation

In the following chapters we will give examples of these types of parallelism and investigate the concept of inherent parallelism further.

Problems

3.1 Show that the hard-sphere problem can be brought into the form of an event-driven simulation algorithm [3.9].

3.2 **Pie Parallel Algorithm:** Consider the Eden model [3.12] for growth of compact clusters. In this model a seed-site is initially occupied. One of the four neighbour sites is randomly set to be occupied in the next step. Again one of the empty neighbouring sites of the occupied site is selected randomly and occupied. This process is repeated until either no computer storage is left or one decides that the cluster is large enough. To parallelize this process of cluster growth we can employ a pie-like decomposition (in two dimensions)

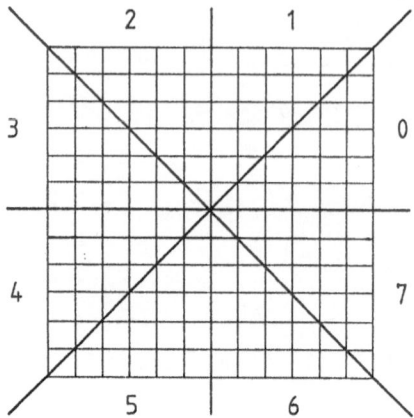

Fig. 3.4. Pie decomposition of a two-dimensional lattice. Each sector of the pie is assigned to one processor

of the underlying lattice (Fig. 3.4). Each of the sectors can be operated upon quasi-independently. Write a computer program for the generation of Eden clusters using the pie decomposition of the lattice. What kind of communication network would you need? What about synchronization?

3.3 Use the above idea of the pie parallel algorithm to develop a program for the diffusion limited aggregation problem.

3.4 Would the pie parallel algorithm idea also be feasible for the Leath [3.13] method of generating large percolation clusters?

4. Concepts of Parallelism

The question which we pose and would ultimately like to answer is whether for a given physical system there is a combination of a formulation of the degrees of freedom and a method for the generation of states that is optimal, in the sense that it has the highest degree of parallelism. This does not imply that it is optimal with respect to the physics, since there may be specific physical grounds why one is inclined to study a certain model or formulation of a model.

Suppose that we have two models each describing the same physical system. The two models have the relaxation time behaviour

$$\tau_1 \sim \xi^{z_1}, \quad \tau_2 \sim \xi^{z_2} \tag{4.1}$$

with $z_1 > 1$ and $z_2 < 1$. (We refer the reader to the dynamic interpretation of simulation methods described in Sect. 2.2.3.) ξ is the relevant physical length scale in the problem. Clearly, as ξ increases, the model with the second relaxation time behaviour will win (Fig. 4.1) since the time correlations are weaker than in the first case and one therefore needs to generate fewer states. There is a break-even point where $\tau_1 = \tau_2$, whose position in terms of the various coupling constants is influenced by the algorithm and its complexity in generating new states.

4.1 Some Basic Definitions

Even though ideal machines are mainly of theoretical interest they serve an important role in helping to understand some basic concepts. Not only do such

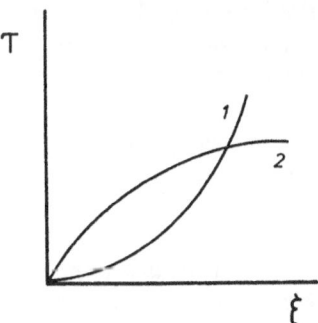

Fig. 4.1. Two relaxation time behaviours for one physical system that is described by two models

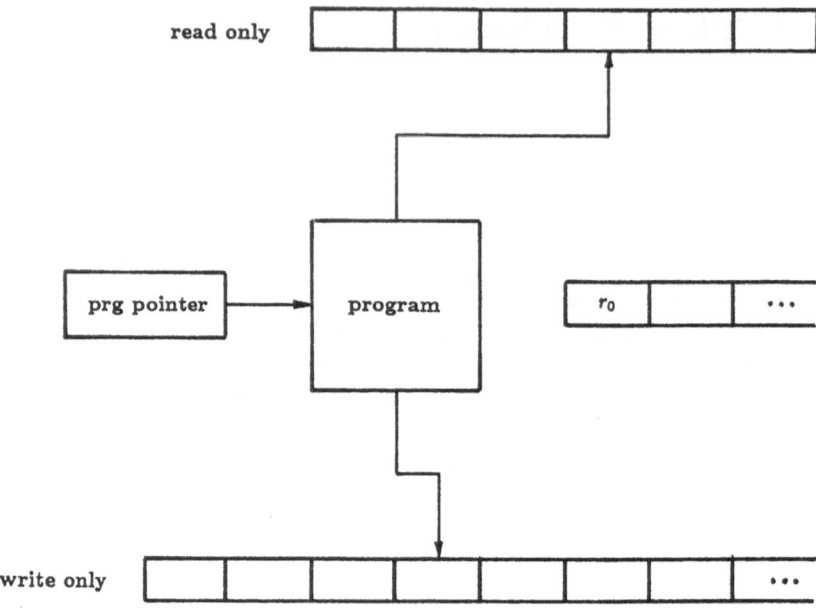

read only

prg pointer → program r_0 ...

write only   ...

Fig. 4.2. Scheme of a random access machine (RAM)

ideal machines influence the design of practical machines (and of course vice versa), they also influence the design of algorithms and their performance. We shall thus state some of definitions pertaining to our task.

Definition 1. *A* **Random Access Machine** *(RAM) [4.1] consists of memory, a read-only input tape, a write-only output tape and a program. Stored on the input tape is a sequence of integers. After each read the input tape moves to the next integer and after each write on the output tape the tape moves to the next vacant space. The memory is composed of an arbitrary number of registers $r_1, r_2, ...,$ which can each hold one integer. With the register r_0 one can perform computations. The program cannot be modified.*

Figure 4.2 shows the scheme of a RAM. The definition tries to capture what is precisely meant by "computing" and describes, more or less, the von Neumann machine and its model of computing. In this context computing is explicitly serial. There are many equivalent definitions and we shall not discuss all their ramifications.

It is more important from our present point of view to specify what is meant by parallel computing. One possible definition of a parallel machine for parallel computing is the following:

Definition 2. *A* **P-RAM** *[4.2] is a machine with an arbitrary number of processors which work synchronously on a program and communicate via a common memory. We assume a constant cost for arithmetic and communication operations.*

During one cycle every processor can read every memory cell. An arbitrary number of processors can read one memory cell, but only one processor is permitted to write to one cell at any moment.

Clearly this is one possible model of parallel computing. Specifically, it entails a model of communication, but circumvents the problem of interprocessor communication. All computations are synchronous, so that homogeneous algorithmic structures are most suitable. Indeed in many numerical applications the data, and hence the algorithmic structures, are homogeneous, in which case there is only one set of operations which need to be carried out on a set of data.

It is more difficult to define models where the mode of execution is not synchronous and the memory is not central to all processors. We will not pursue this theoretical discussion further here, but later we adopt a more practical stance and look at machine designs directly.

4.2 The Complexity of Computation

In the introduction we encountered a problem which did not yield to parallelization and where the complexity was the same for a serial or a parallel machine. This raises the possibility that there are problems in computational science which are hard to parallelize.

The complexity of a problem tells us more or less whether or not we can solve it in a reasonable time. More precisely, it can tell us how efficiently we can solve the problem. Firstly, let us define what we mean by complexity for a single processor machine. The *time complexity* measures the "time" it takes to execute an algorithm of a given problem size n. Time here means the number of operations that have to be executed.

In what follows we assume the same cost function for each basic operation, irrespective of whether it is a multiplication, addition or any other operation, and we assume that each basic operation takes the same number of processor clock cycles to complete.

In general it is possible that the problem size is not constant but varies. A simulation of a model in the grand canonical ensemble furnishes an excellent example of varying problem size, i.e., the number of particles changes. The complexity is then no longer constant but fluctuates. In this case we define the time complexity either by the *worst case time complexity* or by the *expected time complexity* . Continuing with our example of the grand canonical ensemble simulation, we can calculate the time complexity by considering the maximum number of particles which we can simulate or alternatively consider the average number of particles present in the system. If we do not specify explicitly that the time complexity referred to is the expected time complexity then we always mean the worst case complexity. Of less interest at the moment is another measure, the space complexity, whose existence is only mentioned here in passing.

Before proceeding we study the time complexity of the Hamiltonian

$$\mathcal{H} = \frac{1}{2} \sum_i m v_i^2 + \sum_{i<j} V(r_{ij}) \quad , \tag{4.2}$$

where we assume that the potential is long ranged, so that every particle interacts with all other particles in the box. In the mirco-canonical molecular dynamics simulation we must integrate the equations of motion. Since every particle interacts with all other $N-1$ particles, of the order of N^2 operations will be necessary to perform one integration step. To calculate the position and momentun of each particle several multiplications and additions are necessary. Their precise number is not important in the time complexity, since the dominant part is the N^2 interaction calculation. The time complexity for the long-range molecular dynamics in the straightforward algorithm is $O(N^2)$. In quoting the complexity we use the standard big-oh notion [4.3].

On a single processor machine we define an algorithm which has a time complexity that does not grow faster than n^k as *efficient* [4.4]. Any algorithm with time complexity which grows faster than n^k is called *inefficient*.

Let us consider now the complexity of an algorithm for a parallel machine. Analogously to the mono-processor case we define the time complexity (again, we consider only the worst case time complexity) for a given problem size n as the time that elapses from when the first processor starts until the last processor finishes the algorithm [4.1].

A *polylogarithmic time algorithm* takes $O(\log^k n)$ parallel time for some integer k. An *efficient parallel algorithm* takes polylogarithmic time using a polynomial number of processors. The class of problems which can be solved with efficient parallel algorithms is called the *NC class* [4.5]. A subclass of NC are the *optimal parallel algorithms* where the product of the parallel time and the number of processors is linear in the number of processors. This is, of course, the ideal situation that we are striving for, as was indicated in Chap. 1. It is not known what types of simulation problems fall into this class. More research is needed to work out what problems one can bring into a form susceptible to efficient and optimal parallelization.

4.3 More on Models and Methods

From the computational science point of view the parallelism of a system can occur on different levels. We assume that we intuitively understand what is meant by a physical system S. Such a system can have different descriptions. A familiar example already cited is the Ising system, where we can have a description with the entities being the spins on a lattice with the Hamiltonian

$$\mathcal{H} = -J \sum_{\langle ij \rangle} s_i s_j - H \sum_i s_i$$

and the corresponding partition function

$$Z_1 = \sum_s \exp(-\mathcal{H}/k_\mathrm{B}T) \quad ,$$

or a description with the entities being clusters in a percolation-type reformulation (in zero field)

$$Z_2 = \sum_s B(s,T) 2^{c(s)} \quad .$$

A system can be clad with different descriptions and the computational complexity of the generation of a state from the previous state will vary with the description. The degree of parallelism inherent in the description of the physical system will also change.

From the physics point of view we want to investigate a physical system with fixed thermodynamic and dynamic properties. We do not want to change the model simply because of a possible better parallelization. Our aim must therefore be to find the best algorithm under these constraints.

Let $\mathcal{Z} = \{Z | Z \text{ is a partition function}\}$ be the set of all partition functions. On this set we define an equivalence relation as follows. Let Z and Z' be two partition functions. Z and Z' are equivalent if all equilibrium thermodynamic properties are identical in the thermodynamic limit.

Each equivalence class defines a physical system S. Furthermore for every $Z \in \mathcal{Z}$ there exists a description D which describes the entities or the basic constituents for the phase space $\Omega(D)$. Examples of such descriptions are the spins of the Ising model or the Potts variables as compared to a cluster formulation of the same models, which give an alternative description of the same physical system.

We further let $\mathcal{M}(D)$ denote the set of simulational methods for the system with the description D. If $M \in \mathcal{M}(D)$ then M can be thought of as a function operating on the phase space $\Omega(D)$ as follows:

$$x_{i+1} = M(x_i) \quad , \qquad x_i, x_{i+1} \in \Omega(D) \quad . \tag{4.3}$$

The set $\mathcal{M}$ includes Monte Carlo methods, molecular dynamics, hybrid methods and so forth. The subset $\mathcal{MC}(D) \subset \mathcal{M}(D)$ denotes the Monte Carlo methods which one is able to construct, and similarly obvious abbreviations can be used for the other methods. The Glauber function (2.14) and the Metropolis function (2.13) are the best-known examples of Monte Carlo methods.

A model of a system S is defined by the tuple (D, M), where D is a description of the model or a partition function and $M \in \mathcal{M}(D)$ is a method of generating a new configuration from the old one. The examples at the beginning of this section are hence two models of the same system.

If (D, M) and (D, M') are two models derived from the same physical system S, then they are *equivalent with respect to their dynamics* if they have exactly the same asymptotic relaxation times.

Note that we only require the systems to have exactly the same dynamic behaviour in the thermodynamic limit. For finite systems we anticipate possibly different finite-size effects. It may be that amplitudes only match in the thermo-dynamic limit or that some relaxation times are missing and only gradually come into play as the system size is increased.

The complexity of the model (M, D) results then from the complexity of the method M or, better stated, the algorithm by which a new configuration is generated from the previous one.

We can now define what we mean by the inherent parallelism of a model:

Definition 3. *The* **inherent parallelism of a model** $IP(D, M)$ *is defined as the complexity of M on a serial machine divided by the complexity of M on a parallel machine (P-RAM).*

What is the inherent parallelism of the model where the Hamiltonian is given by the Ising model with the spins as entities on a d-dimensional lattice and the Glauber function is used for the transition probabilities? On a serial computer we must visit each spin to produce a new configuration, and one update requires $O(L^d)$ operations. On the parallel computer we can decompose the lattice into two interpenetrating sublattices. On each of the sublattices the spins are independent in the sense that the Glauber function and the nearest neighbour character of the Hamiltonian only require that we know the spins from the other sublattice. The computational complexity is thus $O(2)$ and the inherent parallelism of this model is $L^d/2$.

We can ask two questions. The first is whether for a given description there is an optimal method M, such that for all other methods M'

$$IP(D, M) \geq IP(D, M') \quad ,$$

and secondly, whether there is a pair (D, M), such that for all other pairs (D', M') the following relation holds:

$$IP(D, M) \geq IP(D', M') \quad .$$

Unfortunately we are unable at the moment to present a proof that indeed such optimal descriptions or methods either do or do not exist.

From a more practical point of view we would like to modify the above definition of inherent parallelism to accommodate the fact that all real-life computers have only a limited number of processors and a specific interprocessor communication network. Such a computer is definitely not able, in general, to sustain the theoretical complexity of an algorithm. Some compromise must be made as to the number of processors, problem size and space. We come back to this point in Chap. 5, where we consider the matching problem.

4.4 Performance Measurements

The performance of an algorithm and its measurement are important but elusive issues. Ultimately we want the job to be finished in less (wall clock) time when we invest more computing power, i.e., employ more and more processors.

On serial computers the problem of evaluating how well an algorithm performs is already a complicated matter. On a parallel machine the situation is even more complex due to two new factors. Firstly there is the communication between processors, and secondly there are situations where some processors are left idle at times but we gain in the overall performance.

In the introduction we declared that our goal is to develop algorithms which are in some sense efficient. One of the criteria for efficiency is the scaling behaviour with respect to the problem size n. In general the algorithm is also dependent on the number of processors p employed. In the case of the theoretical machines discussed above this is not necessarily so, since it is always assumed that the number of processors is sufficiently large to sustain the algorithm.

Let us define the scaling behaviour with respect to the number of processors used in a particular algorithm $\mathcal{A}$ as

$$S_{\mathcal{A}}(p) = \frac{\text{time complexity on one processor}}{\text{time complexity on p processors}} \; . \tag{4.4}$$

This notion of the *speed-up* [4.6] gained over one processor is of course not without problems. First and foremost, an algorithm on one processor will certainly look different from one which is specifically designed to run with more than one processor. Secondly, the communication cost which certainly enters for most algorithms once the problem is parallelized is completely disregarded here. Start-up times and latency for communication on real machines, which can greatly influence performance, are not taken into account. As the definition stands the performance will look different on machines with dissimilar architecture. In the above definition the problem size dependence is also completely lost.

Before proceeding further we define the *utilization* of a p-processor machine as

$$\eta_{\mathcal{A}} = \frac{S(p)}{p} \; . \tag{4.5}$$

From the remarks at the beginning of this section it is clear that utilization of all processors does not necessarily mean that the job is finished faster in wall clock time than with fewer processors!

Both definitions, the speed-up and the efficiency, are machine dependent. They make a statement about the algorithm/machine pair and not about the algorithm itself. There are, however, performance characteristics which depend not on the actual machine, i.e., on the number of processors used, but on the *type* of architecture. We will follow up this point when we discuss the performance of geometrically parallelizable problems.

Problems

4.1 **Multi-Spin-Coding:** [4.7–9] With multi-spin-coding one can achieve parallelism on the bit level, even on a serial machine. What is the time complexity of the two-dimensional Ising model, using the Glauber function for the Monte Carlo process, applying Multi-Spin-Coding on a serial machine?

4.2 **Cellular Automata:** [4.10] Cellular automata are not just another way of simulating the Ising model. Their application reaches into many areas of science. For simple one-dimensional cellular automata calculate the time complexity for a synchronous and an asynchronous update.

4.3 **Creutz Algorithm:** [4.11] The Creutz algorithm is one way of introducing a constant energy into an Ising model simulation. In its simplest form there is one demon which can move across the lattice and flip spins. The demon can carry energy. If the spin flip results in a loss in overall energy, then the demon collects the energy difference. If the system would gain energy by the flip, then the move can only be made if the demon carries enough energy. Suppose now that there are m demons which can move across the Ising lattice (L^d). Calculate the time complexity. Can you speed up the calculation? (Hint: Multi-???-Coding)

5. Parallel Machines and Languages

From the point of view of computational or simulational science there simply cannot be too great a variety of computer architectures. The diversity of problems being investigated is enormous, including for example the inversion of large matrices, as in Monte Carlo simulations of lattice gauge theory; problems in molecular dynamics, where the interaction between particles can be short range or else one needs to calculate the interactions between all pairs of particles; and Ising-model types of Monte Carlo simulations for statistical mechanics systems. In principle each of the problems can be dealt with on any architecture. However, in practice not every architecture fits the needs of the problem. In some situations where the problem being investigated and the computer architecture are particularly ill-matched, the performance is degraded so much that a significantly less powerful computer that has an appropriate architecture can achieve a performance that is an order of magnitude better.

5.1 General Purpose Parallel Computers

Theoretical models of parallel computing illustrate a number of possible parallel computer architectures, but not all of these have physical realizations. It is hardly possible, for example, to realize a computer architecture with an unlimited connectivity. The connectivity is limited in practice to k input/output channels. Hence complexity measures for problems need to be appropriately modified, e.g., $O(\log n)$ becomes modified to $O(\log_k n)$.

The difficulty in presenting the possible and physically realized architectures in a concise way is shown by the diagram in Fig. 5.1, where we illustrate the basic features that a parallel computer can have. These features include the number of processors, their mode of operation, the memory organization and the connectivity between the processors. The number of possible combinations is quite large, and in this chapter we concentrate upon some realizations in existing computers.

5.1.1 Processor Concepts

There are two main philosophies behind the architecture of most existing parallel machines, excluding some very exotic ones. On the one hand there are machines with a small number of very powerful processors, whose performance

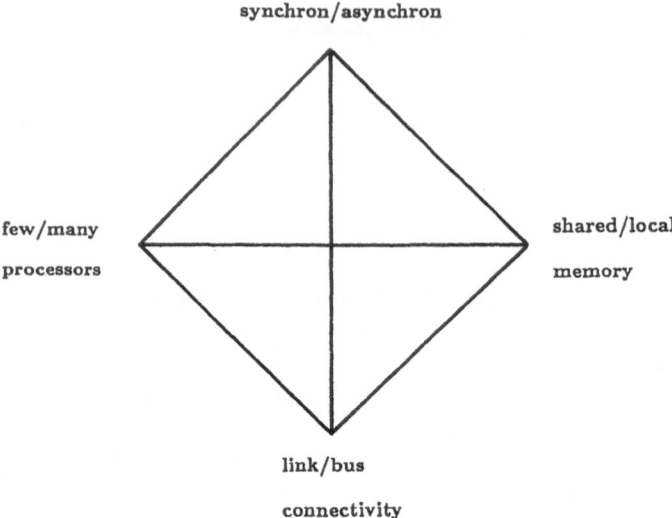

synchron/asynchron

few/many
processors

shared/local
memory

link/bus

connectivity

Fig. 5.1. Some possible combinations of the processing elements for a parallel computer

with respect to the number of instructions per second, for floating point as well
as integer operations, are in the range of more than 100 MOPS (Mega Operations Per Second; an operation can be an instruction, an integer add or a floating
point multiply). On the other hand there are machines with a very large number
of processors, each of which is much less powerful, and indeed some of which
have only bit-level capabilities. The performance of a single processor typically
lies well below 100 MOPS.

In the first category fall machines like the Cray [5.1], Alliant [5.2] or Convex
[5.3]. These machines have only two, four or eight processors, each in itself a
general purpose processor organized more or less according to the von Neumann
principle. Attached to each of the processors is a vector unit, so that they really
represent a combination of two further classes of machines.

In the second category are machines like the ICL DAP [5.4] and the Connection Machine [5.5], which have processors on the bit level, the Meiko Computing
Surface [5.6] and the Supercluster [5.7], the last two of which are based on the
transputer and therefore have a larger granularity , i.e., the power of each processor is greater. Each transputer is itself a general purpose processor, although
the number of instructions or arithmetic operations which it can perform is considerably less than one processor of the Cray or any of the other machines in the
first category! Such a design facilitates heterogeneous data structures whereas
machines like the Connection Machine favour homogeneous data structures, and
therefore have a close similarity to the vector machines in the first category.

There are also machines which fall between the two categories. One such
machine is the Suprenum computer [5.8]. This machine has many intermediate-
scale processors and attached to each of the processors is a vector unit.

One useful means of categorizing the various types of architectures is as follows:

- SIMD: Single Instruction Multiple Data

- MIMD: Multiple Instructions Multiple Data

From the point of view of computational science these two architecture types essentially pose the following questions:

- SIMD: Can the problem be brought into a vector form?

- MIMD: Can the problem be partitioned?

Most of the machines listed above follow the concept of an asynchronous mode of operation in which each processor can operate independently of the others. For the machines in the first category this means that each of the powerful processors can operate more or less as a conventional computer independent of the other processors. This is also true for the Computing Surface and the Suprenum machine. They are therefore classified as MIMD computers.

A synchronous mode of operation is displayed in computers such as the DAP and the Connection Machine. Since each processor is following both the same clock and the same instruction they are classified as SIMD machines. Here a homogeneous data structure is favoured, as in the case of vector processors.

In a parallel computer messages can either be passed via a communication network or by accessing a common memory. If the memory is local to the processor (e.g., the Computing Surface, Hypercube, etc.) messages need to be exchanged via a network. All requests for non-local memory are handled by the associated processor.

A taxonomy of the possible machines is extremely difficult. Many attempts at classification have been made [5.9–11] and here is not the right place to make another attempt. One of the principal difficulties of such an undertaking is the wide variety in the connectivity of the machines.

5.1.2 Communication Networks

The traditional way to exchange messages is via a shared memory (e.g., Cray, Convex, etc.). A shared memory machine comes close to the theoretical parallel computer discussed in the previous chapter. All processors can use the shared memory for storage or message exchange, but precautions must be taken to avoid simultaneously reading or writing to the same memory location.

While the connectivity of the processors in machines with few processors is straightforward, many different possible solutions exist for massively parallel machines. The connectivity ranges from solutions where a bus is used, as in the Suprenum machine, to butterfly networks, as in the Butterfly [5.12], hybercubes, in the Hypercube [5.13,14], and variable topologies in other machines.

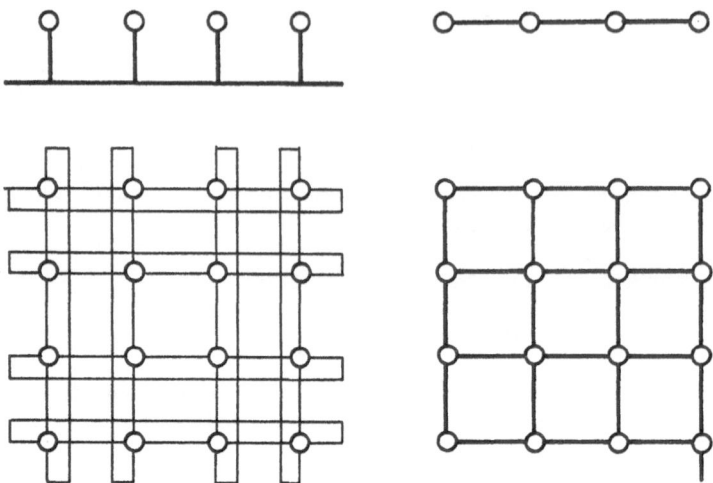

Fig. 5.2. Bus-type (*left*) and link-type (*right*) communication networks for parallel computers

The connectivity is greatly influenced by the communication model. From a general viewpoint we distinguish between bus and link communication networks. The difference becomes apparent in Fig. 5.2. While a message in a bus-oriented network can be broadcast directly to all processors, a routing through other processors must take place in link-oriented networks. In link networks only neighbours can talk directly to each other.

A communication network tries to enable a direct connection between as many processors as possible. A general communication network realizes the mapping

$$N \rightarrow M \quad , \tag{5.1}$$

where N denotes the entire set of processors and $M \subset N$ is a subset of the processors. In the case $N = M$ every processor is connected to every other processor. This is not to be confused with a bus, along which every processor can communicate with every other processor attached to the bus. What is meant here are private lines between the processors. A crossbar realized by a bus would present a bandwidth bottleneck, since the bandwidth of a bus cannot be made arbitrarily large. If all processors communicated at the same time the bus would be overloaded. However, with local communication this bottleneck is removed, although generally at the expense of restricted connectivity.

A *crossbar* communication network is shown in Fig. 5.3. All communication graphs can be generated with such an underlying crossbar structure, although of course such a network is technologically possible only for a small number of processors because of the prohibitively large number of connections (N^M). The number of possible connections must therefore be restricted, and there are two possible routes to follow. We can either fix the topology of the communi-

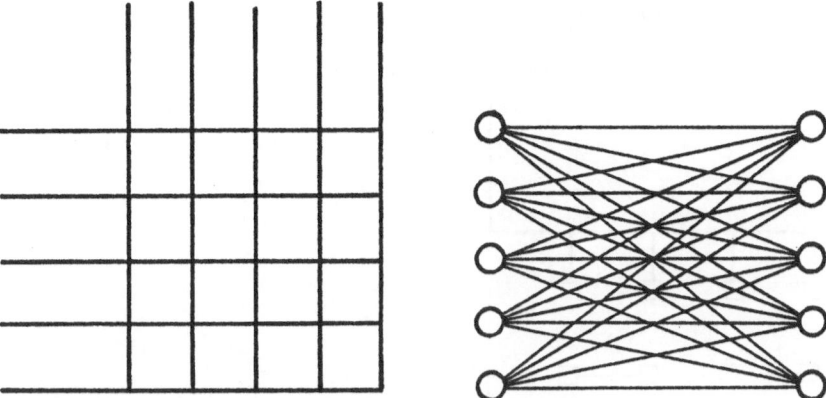

Fig. 5.3. Different plots of the crossbar communication network

cation network or we can arrange a network which is able to admit as many communication graphs as possible.

Of the N^M possible connections of the general communication network only the $N!$ one-to-one mappings are of interest. For a fixed neighbour relation we can have

- a ring

- a lattice

- trees

- a hypercube

or topologies that are either structurally similar or made up of composites of these.

The *ring* structure is possibly the simplest to realize (Fig. 5.4) and can be accomplished with either single interprocessor links or a bus. The same is true for the lattice topology, and this class of networks is widely used for algorithms in simulational science. Almost all geometrically parallelizable problems can be mapped onto such topologies.

Consider now the path length that a message must travel between two processors for such local communication networks. If the two processors are not nearest neighbours then a routing of the message via intermediate processors must take place. For the ring network the path length grows linearly $O(N)$ with the number of processors in the ring. The length of the path for the lattice network is only $O(\sqrt{N})$, and for the tree it is even smaller $O(\log N)$.

Shorter communication paths exist in *hypercube* networks (Fig. 5.5). A hypercube network consists of $n = 2^k$ nodes, and each node can be conveniently identified by numbering them $0, ..., 2^k - 1$. Two nodes are adjacent if their respective labels differ by exactly one bit position.

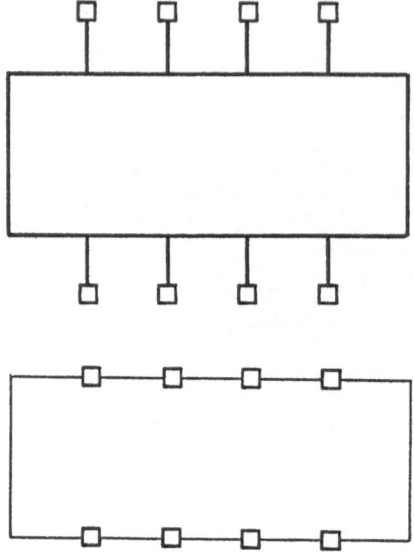

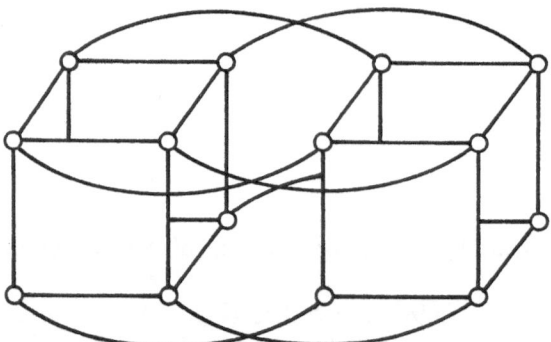

Fig. 5.4. Ring communication network with a bus and links

Fig. 5.5. Hypercube communication network

In addition to these communication networks with a fixed neighbour relation we can have networks with variable neighbour relations, such as

• exchange or permutation nets
• mixtures between crossbar and exchange

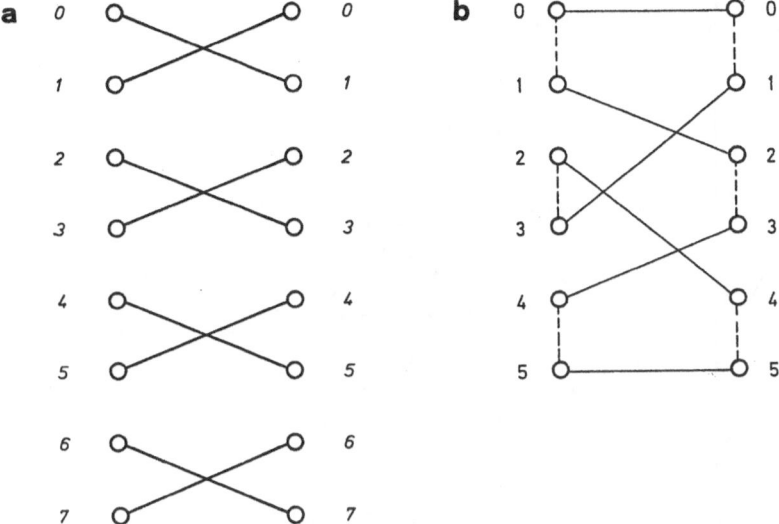

Fig. 5.6. Exchange communication networks: (a) permutation exchange network; (b) shuffle exchange network

Such networks, and especially the second set, allow the embedding of many communication graphs. The Meiko Computing Surface and the Parsytec Supercluster have the possibility of electronically configuring the network according to the needs of the program. Which processor is connected directly to which other processor is not predetermined but rather configured according to the communication graph appropriate to the program. The graph does not need to have a simple topology, such as the ring, and can be very irregular.

In the following discussion of exchange networks we assume for convenience that the processors i are labelled in a binary representation:

$$i \in \mathcal{N} \rightarrow (b_k, ..., b_1) \in \mathcal{Z}_2^k \quad , \tag{5.2}$$

where $\mathcal{N}$ denotes the set of natural numbers and $\mathcal{Z}_2$ the set $\{0, 1\}$. The first such network is the *exchange permutation*, where two processors always exchange their messages. Formally this is written as

$$\pi_e^{(\iota)} : (b_k, ..., b_\iota, ...b_1) \rightarrow (b_k, ..., \bar{b}_\iota, ...b_1) \quad , \tag{5.3}$$

where $\bar{b}_\iota$ denotes the binary complement. ι determines how far away the exchange takes place. An example of such an exchange network is displayed in Fig. 5.6a.

The *shuffle exchange* network is slightly more complicated. Formally we write

$$\pi_s : (b_k, ..., b_1) \rightarrow (b_{k-1}, b_{k-2}, ..., b_1, b_k) \quad . \tag{5.4}$$

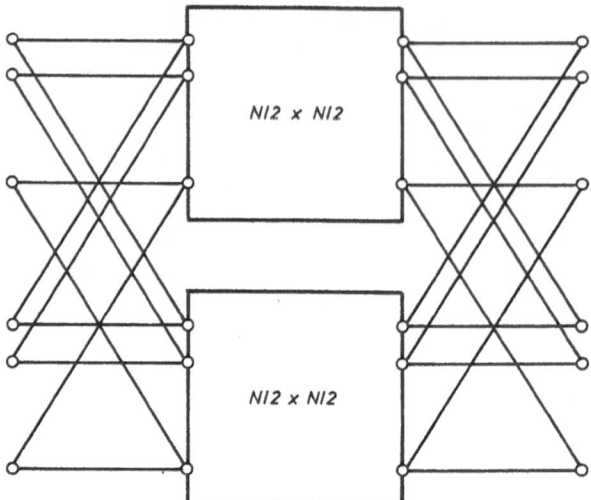

Fig. 5.7. Benes exchange communication network

Informally the perfect shuffle is indeed like the shuffling of a deck of cards. We first cut the deck in two halves, then interleave the cards such that the first card of the top half comes on top of the first card from the second half. The second card of the first half comes next and on top of the second from the second half, and so on. The connections in a perfect shuffle for four processors are shown in Fig. 5.6b.

As a last example of an exchange network we discuss the *Benes exchange network* [5.15]. Benes showed that a $N \times N$ crossbar can be reduced to a $N/2 \times N/2$ crossbar together with an N input exchange switching network. This network is illustrated in Fig. 5.7. Such a network admits a large number of communication graphs and also allows the possibility of an electronically configurable network.

5.2 Parallel Machines for Special Physics Problems

For the numerical investigation and solution of particular physics problems there are good reasons not to use the general purpose computers described above, but rather to build *special purpose machines* which match either the problem itself or the particular algorithm used to solve the problem. Perhaps the most obvious and important reason for building dedicated machines is to secure time and availability of the substantial computing power required.

For some problems, such as lattice gauge theory calculations or many statistical physics problems, the computer time required for the solution is prohibitively large for conventional computers. For example, investigations of the spin glass problem [5.16] on a special purpose computer used the equivalent of one year of

Cray time [5.17]. Moreover the price of building a special purpose machine may be much less than a conventional one. To construct such a machine it is sufficient to use silicon chips, which are both readily available and inexpensive (i.e., no new technology is required), so that the cost–performance ratio is favourable.

Beside these more commercial arguments there is also the possibility of very large increases in performance. By carefully considering the structure of the problem with respect to

- data

- algorithm

- communication structure

it is possible to tune the machine to the needs of the problem being investigated. It is then further possible to optimize one or more of the following features:

- storage organization

- processor organization

- connectivity

- certain types of arithmetic operations

so that the machine is tailored to match, to as large an extent as possible, the problem and algorithm. The disadvantage of a special purpose machine lies in its inflexibility. New techniques and algorithms, often apparently only slightly different, or developments in the types of questions we wish to answer about the system under study may render the machine ineffective.

In the following we are going to look at a selection of special purpose machines that have been built. We will look at two Monte Carlo machines for statistical physics problems and also at a machine for the molecular dynamics simulation of N-particle problems.

5.2.1 Monte Carlo Machines

A machine built for the Monte Carlo simulation of the generic Ising model illustrates some of the strategies discussed above which lead to an increase in the speed with which simulations can be carried out. We recall here the Monte Carlo algorithm for the simulation of the Ising model with the Hamiltonian

$$\mathcal{H} = -J \sum_{\langle ij \rangle} s_i s_j - H \sum_i s_i \quad , \tag{5.5}$$

where s_i are spins which can take on the values ± 1, J is the exchange coupling between the nearest neighbour spins and H is an external magnetic field.

The algorithmic structure of a Monte Carlo simulation [5.18–22] is as follows:

Algorithm: Monte Carlo algorithm for the Ising model

---

    start from a random configuration s_0
    FOR *step* := 1 **TO** *max.steps*
        WHILE *still.lattice.sites* **DO**
            choose a new lattice site (with spin s)
            new trial value of spin at the site is $-s$
            compute change of energy $\Delta\mathcal{H}$
            accept trial value or leave the spin as it is
        ENDDO
    ENDFOR

Starting from a spin configuration, e.g., a completely ordered configuration of all spins up (+1) or down (−1), new configurations are generated by going through the lattice and changing the spin orientations. One can go through the lattice either randomly or regularly. A regular update pattern through the lattice is more amenable to fast calculations. Depending on the energy change caused by a reversal of the spin being considered, a decision is made whether to accept the flipped spin or not.

a) The Delft Monte Carlo Machine

The construction of the Delft machine [5.23] (Delft Ising System Processor: DISP) reflects the structure of the Monte Carlo algorithm in a direct way. The Delft machine was built for the simulation of a generalized Ising model with the Hamiltonian

$$\mathcal{H} = -J\sum_{\langle ij\rangle} s_i s_j \quad - \quad J_1 \sum_{\text{nn}} s_i s_j$$
$$- \quad J_2 \sum_{\text{nnn}} s_i s_j - J_3 \sum_{\text{tr}} s_i s_j s_k$$
$$- \quad J_4 \sum_{\text{sq}} s_i s_j s_k s_l - H\sum_i s_i \quad . \tag{5.6}$$

The J_i's are additional exchange couplings for the next-nearest (nn), next-next-nearest (nnn) neighbours, elementary triangles (tr) in the case of a two-dimensional triangular lattice and squares (sq) in the case of a simple two-dimensional lattice.

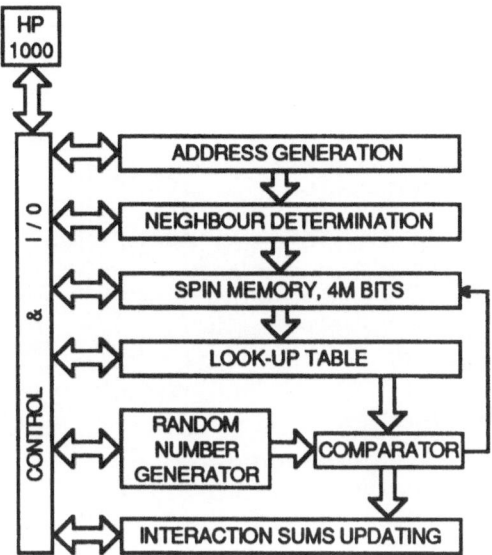

Fig. 5.8. The functional design of the Delft Monte Carlo machine for Ising-type problems

As well as the two-dimensional lattice the machine is able to simulate simple cubic and face-centred-cubic lattices and has a memory capacity of 2^{22} spins. The functional construction of the machine is shown in Fig. 5.8. The design reflects the same structure as the Monte Carlo algorithm. In the first step a spin is selected for a possible spin reversal. Since the neighbouring spins (the exact number of which depends upon the Hamiltonian under consideration) are needed for the energy calculation, the next step consists of retrieving their values from memory. The transition probability for a change is looked up in a table, which is drawn up in advance since there are only a small number of possible transition probabilities. While all of these steps are being executed a random number is generated. The generation of a random number is typically the slowest part of the Monte Carlo algorithm, but since it is done in parallel with the calculation of the transition probability the entire process is very efficient. The mapping between the algorithm and the machine is one-to-one and hence the parallelism is extremely functional.

b) The Santa Barbara Ising Computer

The design of the Santa Barbara machine differs from the Delft computer in that it is oriented around the data structure of the lattice. An efficient implementation of the Monte Carlo algorithm requires a continuous flow of spins to generate new configurations. On a simple cubic lattice, for example, a spin update requires the six nearest neighbours (for the simple Ising Hamiltonian) to decide upon the new orientation of the single spin. To guarantee a continuous flow of spins the usual periodic boundary conditions are changed to toroidal, as illustrated in

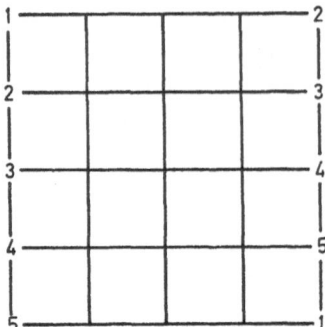

Fig. 5.9. Toroidal or skewed boundary conditions

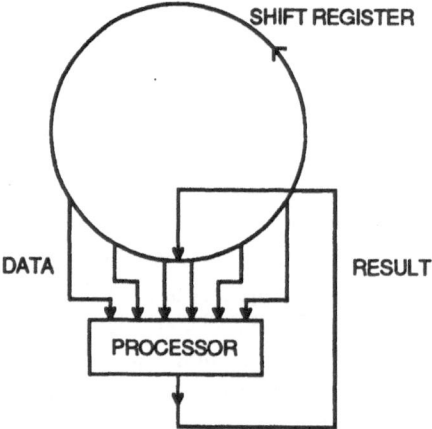

Fig. 5.10. Functional design of the Santa Barbara machine for the Monte Carlo simulation of the Ising model

Fig. 5.9. The finite-size effects of the simulation are then altered by the change in boundary conditions [5.19], but such a change can be dealt with. Indeed it has now become usual practice to change from periodic to toroidal boundary conditions when using vector computers.

The transformation of a three-dimensional lattice into a one-dimensional structure allows us to obtain the neighbouring spins at a constant separation. This is shown schematically in Fig. 5.10.

The architecture of the Santa Barabara machine exploits the inherent parallelism of Monte Carlo Ising simulations, which results from the data structure and the condition of detailed balance. Instead of using just a single processor it is possible to include many more, so that spins can be updated in parallel. The processor thus reflects, in a manner similar to that of the Delft computer, the structure of the Monte Carlo algorithm. The Santa Barbara machine optimizes the performance by exploiting both the data structure and the algorithmic structure.

For some orientation regarding the speeds of these special purpose computers let us compare them with speeds obtained using a commercially available computer. The best speed on the Ising problem for the DISP was 1.5×10^6 spin updates per second and for the Santa Barbara machine 2.5×10^7. Using the ICL DAP a speed of 2.7×10^6 updates per second was obtained [5.25].

As well as these machines for the Ising model, quite a number of other special purpose computers [5.26–31] have been constructed or are under construction for a variety of other physics problems which involve using the Monte Carlo method.

5.2.2 Molecular Dynamics Computers

In this section we consider some of the machines specifically designed for molecular dynamics simulations. The numerical problem consists of solving Newton's equations

$$\frac{dp_i}{dt} = -\sum_{j\neq i}^{N} \nabla_{r_i} V(r_{ij}) \tag{5.7}$$

for a system of N particles with a Hamilton function

$$\mathcal{H} = \frac{1}{2}\sum_i mv_i^2 + \sum_{i<j} V(r_{ij}) \tag{5.8}$$

in its discretized form [5.20,32,33]. Here m is the mass of a particle with velocity v_i at the position r_i. We assume a central two-body potential.

By far the largest fraction of the required computer time goes into the computation of the forces for the $N(N-1)/2$ possible interactions. The naive complexity of the problem is of the order $O(N^2)$. All practical algorithms and special purpose machines aim to reduce this complexity.

In general the forces are short ranged, for example the Lennard-Jones potential

$$V(r) = 4\epsilon\left[\left(\frac{\sigma}{r}\right)^{12} - \left(\frac{\sigma}{r}\right)^6\right] \tag{5.9}$$

with a cutoff at 2.5σ, and thus most of the $N(N-1)/2$ terms are zero. Both algorithms and specially designed machines exploit this feature. Two particles which are far apart cannot possibly influence each other and their interaction can therefore a priori be eliminated from the force calculation. There are many schemes based on this idea [5.32–35], some of which are discussed in Sect. 7.4.

The best-known special purpose machines are those of the Delft group [5.36] and the IBM Almaden group [5.37,38]. There are also a number of new and interesting proposals [5.39].

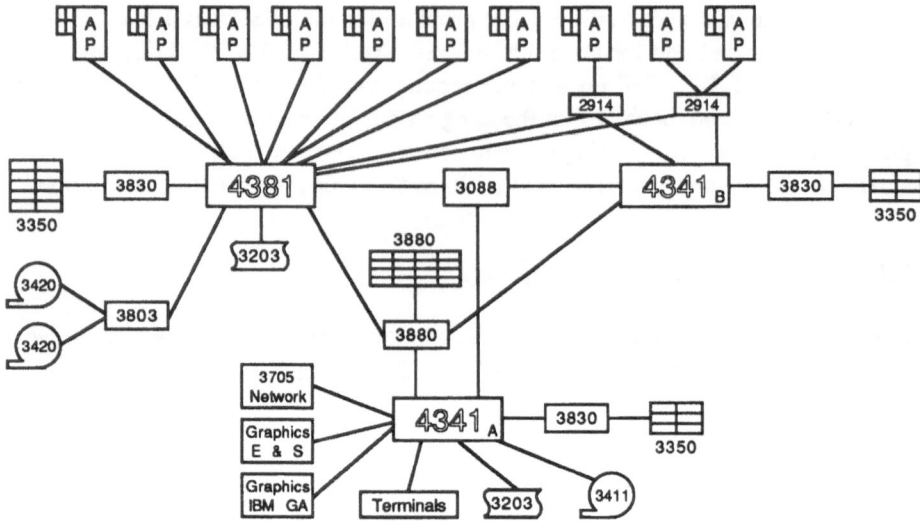

Fig. 5.11. Functional design of the IBM Almaden Research Center machine for molecular dynamics simulations

a) The IBM Almaden Research Center Machine

The idea that is used in the IBM Almaden Research Center machine [5.37,38] is in essence the same as in many software algorithms. To reduce the complexity of the problem the three-dimensional space is partitioned into cells. Each particle interacts (we assume that the forces are sufficiently short ranged) with the particles inside the same cell. In addition it is possible that each particle interacts with the particles in the eight neighbouring cells (in a two-dimensional problem). The cell size must therefore be at least as large as the cutoff distance of the potential.

In this design each processor is assigned one cell. The forces inside the cell are calculated and information has to be exchanged with the neighbouring processors to complete the force calculation for each particle inside a cell. If the problem is large enough then the order $O(N^2)$ is reduced to a linear order $O(N)$.

Figure 5.11 shows the design of the machine. It is apparent that this construction is not specific to molecular dynamics simulations but rather that all problems which can be geometrically parallelized can in principle be run on this machine. The necessary exchange of messages is done via a Benes permutation net.

5.3 Languages for Parallel Computers

Both machine architectures and computer languages have considerable influence on the way we perceive problems and formulate algorithms to solve them numerically. In turn the architectural design of a machine favours algorithmic constructions that appear in the language which is most sympathetic to these hardware features. It is therefore astonishing to see how resilient old computer languages are, even though they can only partially cope with the new world of parallel computing. Indeed, in the worst case they can even prove to be an obstacle to parallel computing.

There are two basic approaches to languages for parallel computing. Perhaps the least inspiring, though possibly most practical approach (especially for the dusty deck problem), is to augment existing languages with calls to subroutines which then handle all the message transfers, the creation of processes and so forth that are necessary for parallel processing. Subroutine augmented languages retain their essential character, together with all their shortcomings, and can offer the new concepts only in a rather restricted way. This is essentially the same idea as that of directly augmenting an existing language by constructs designed for parallel computing.

For vector machines with only one or very few processors Fortran is the main language used in scientific computations. This language has always been a favourite for numerical computation, despite its many shortcomings, e.g., the limited data structures which the language offers.

One- or few-processor vector machines are specifically designed to accelerate vector-type operations. The most obvious example of this is the matrix-vector multiplication

$$Ax = b \quad , \tag{5.10}$$

where x and b are two n-component vectors and A is an $n \times n$ matrix. On vector machines this operation can be expressed in the following schematic form:

```
      do   10   i=1,n
        s = 0.0
        do   20   j=1,n
          s = s + a(i,j) * x(j)
20      continue
        b(i) = s
10    continue
```

All algorithms wanting to exploit this type of vector acceleration must use these basic kinds of operations on data structures that are vector-like.

In advanced forms of parallel languages it is possible to express the operation on a data structure directly as in (5.10). The compiler must resolve the operation and distribute the task to the available processors. This raises the question of automatic vectorization or automatic parallelizers.

Automatic vectorization can be handled today quite efficiently. Not only can loops of the simple sort presented above be vectorized, but also loops with conditionals within their range. Such a vectorization is only passive, in the sense that the vectorizer looks for certain types of constructs and resolves these for processing on the vector units. There is no semantic vectorization.

Similarly, automatic parallelization can only be done on the most trivial level. Certain kinds of "do" loops can be split and distributed among the processors. This is a parallelization not on the operations level, as is the case for vectorization, but on the level of constructs. However, it must be stressed that at present, and perhaps for some time to come, automatic parallelization will be not as effective or efficient as automatic vectorization.

So much for the computational elements of the language. What about communication? A central part of a parallel program is, after all, the communication. Neither Fortran 77 [5.40] nor Fortran versions on the horizon have extensions for communication. The problem lies partly in the many different philosophies concerning the architecture of parallel machines. There is the message-passing type of machine with asynchronous message types on the one hand, and on the other hand there are machines with communication via a shared memory which operate synchronously. The task of finding standard protocols that cover all the various possibilities is an extremely difficult one.

Even for a language such as Fortran there are many different parallel variants. The Fortran versions for the Cyber 205 or ETA, for example, are significantly different from the Fortran used on the Suprenum machine. The Suprenum Fortran [5.41] has language extensions for message passing between processes on different processors, whereas no such extensions exist for the Cyber 205 machine. The reason for this lies in the different organization of the memory. In the Suprenum computer and the transputer-based machines the memory is local to a single processor, whereas there is a large shared memory in the Cray type of machines. There is also the possibility of building the message-passing facility into the language via calls to subroutines, which is the case for the Hypercube Fortran language.

Parallel versions of other languages such as C and Pascal follow essentially the same lines as discussed above, and there are corresponding C dialects which incorporate constructions for parallel computing [5.42].

Languages which are specifically designed for parallel computers are rare and their acceptance is still low. Perhaps the only language which is has been fairly well accepted is Occam [5.43–46]. Occam is the natural language of the transputer and it is based upon message passing. It forms the central subject of Chap. 9, in which a number of features of the language will be presented and examined.

5.4 The Matching Problem

The matching problem has so far only been considered in Sect. 5.2, concerning special purpose machines. The hardware to run the simulations to solve one given problem is typically designed to yield the maximum efficiency with the available

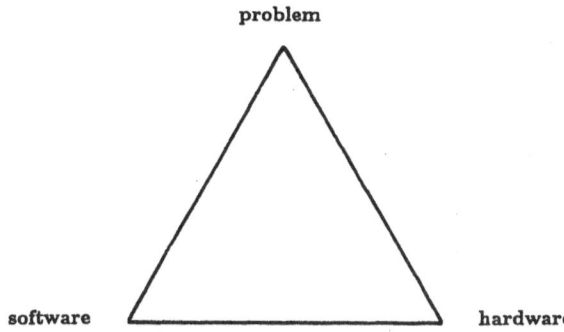

Fig. 5.12. The matching triangle

technology. The software issue was given no attention at all, since for a special
purpose machine it is possible either to hardwire the algorithm or to program the
machine in assembler language or even in binary code. From this standpoint it
is only the relationship between the problem and the hardware that is important.
Let us review the question of matching from a more general point of view.

The matching triangle, as depicted in Fig. 5.12, shows that for an efficient
solution one needs to consider the relation of the problem to the hardware and
to the software and, further, the relation between hardware and software. What
does efficient mean in this context?

• We don't want to spend the rest of our days figuring out how a machine
operates or how to program the simulation algorithm.

• We want to exploit possible the inherent parallelism of the problem, the machine and the language as much as possible.

An almost perfect match of a problem, the machine and the language is perhaps the simulation of the two-dimensional Ising model on the DAP [5.4,47]
programmed in Fortran. The two-dimensional Ising model has an inherent parallelism of $L^2/2$, since half of the spins can be updated in a single move in a Monte
Carlo simulation. The DAP has a two-dimensional simple square processor lattice structure and supports bit arithmetic. This structure corresponds directly to
the simulated system, as shown in Fig. 5.13. The match between the problem and
the hardware could hardly be better. However, there are other machines, e.g., the
Connection Machine, which would in principle fit equally well.

From the point of view of the programming language not everything can
be most efficiently fitted into the frame of Fortran. Here, as in the case of the
architecture, one can in principle express almost any idea in a particular language.
However, if the problem requires a data structure which is not present in the
language it requires considerable effort either to reformulate the problem or to
construct the data structure with the tools available in the language. In both cases
the program will run with less efficiency than when formulated in an appropriate
language.

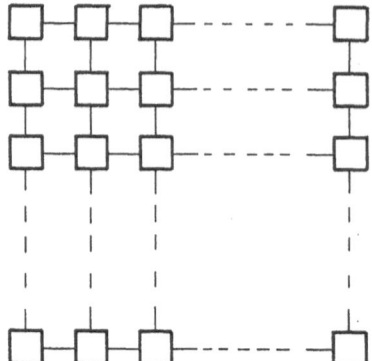

Fig. 5.13. Processor connectivity design of the DAP and the two-dimensional Ising lattice

Let us return to the case of the Ising model. We saw that the most efficient implementation on a general purpose computer could be done on an array processor. One can still exploit part of the inherent parallelism on architectures which do not directly correspond to the problem.

On a serial machine we can exploit part of the parallelism by recognizing that it is possible to store several spins in a single computer word. Logical operators such as "and", "or" and "exclusive or" operate on the entire set of bits in a computer word. It is thus possible, even on serial computers, to treat several spins at the same time. This idea, which was first seen to be possible in 1970 [5.48], is called Multi-Spin-Coding [5.49,50].

On vector machines we can exploit part of the inherent parallelism by considering the data structure of the lattice. The first possible solution is to decompose the lattice into black and white interpenetrating lattices [5.51]. The second possible solution is to label and rearrange the lattice differently [5.52]. In both cases many spins can be treated simultaneously. Likewise it is possible to use the multi-spin-coding algorithm to exploit even more of the inherent parallelism on vector machines.

Needless to say, on coarse grained MIMD machines such as transputer machines or the hypercube, we can split up the lattice among the processors and operate on them simultaneously (see Chap. 7 for a detailed exposition of what is required for such types of algorithms). On the Hypercube and other similar architectures with vector arithmetic units we can para-vectorize the algorithm, as well of course as using multi-spin-coding.

From this discussion we can conclude that even serial machines admit some form of parallelization, at least partially. Of course the concept of the *admitted parallelization* is difficult to define. Many factors influence the possible parallelization. Firstly there is the machine design itself. Almost all serial machines already have some form of parallelism built in on the micro level, extending from instruction prefetch to parallel working I/O units. On the other hand, we must also consider the language the algorithm is programmed in. Some languages do not allow, for example, operations on the bit level. Thus although the machine may admit some form of parallelism, we may not necessarily be able to exploit it.

In the discussion of the matching so far we have neglected the connectivity of the machine. The problem arises, for example, when considering the problem of a simulation of a periodic coarse lattice in two dimensions. Suppose we are forced to work with processors with a coordination number of four. It seems that this is well suited to the lattice problem, since the two-dimensional lattice has indeed the coordination number of four. However, we must not forget that at least one link of one processor is needed to communicate results to the outside world!

If the connectivity does not match the connectivity of the algorithm one is obliged to introduce some routing of messages. In a slightly different context this is similar to the mapping problem [5.53–55]. In the mapping problem there are more processes than there are processors. There is a given communication structure between the processes and a given communication structure between the processors. Since there are more processes than processors one needs to assign several processes to one processor and introduce message routing between the processors and processes.

Problems

5.1 Invent an algorithm which realizes a crossbar (see Chap. 9).

5.2 For a hypercube communication network, what is the path length complexity for a message exchange?

5.3 Show that sorting of N elements takes at least $O(\sqrt{N})$ on a lattice connected machine.

5.4 Assume that $N = 2^k = L^2$, where $k, L \in \mathcal{N}$. Show that the addition of N numbers, which are distributed on the processors with one value per processor, takes

- $O(\log N)$, on a cube connected machine
- $O(\log N)$, on a perfect shuffle machine
- $O(\sqrt{N})$, on a lattice connected machine

5.5 Construct a look-up table for the transition probabilities for the Hamiltonian

$$\mathcal{H} = -J \sum_{\langle ij \rangle} s_i s_j - J_1 \sum_{nn} s_i s_j \quad ,$$

where nn denotes next-nearest neighbours. Can you avoid bank conflicts in the memory?

5.6 For a coarse grain checker-board two-dimensional lattice construct a mapping of the blocks of spins to processors (the processors have a coordination number of four) such that communication to the outside world is also possible.

6. Replication Algorithms

In Monte Carlo simulations as in other computational methods one typically needs to run an algorithm with several different parameters. Consider the example of the two-dimensional Ising model. To determine the order parameter m as a function of the lattice size L and temperature T, i.e., $m(T, L)$, many simulations need to be carried out. Often the lattice sizes we are interested in are fairly small, and so it does not make sense to spread them over many processors. The computational requirement of a single run with respect to storage and time is small enough to allow each individual simulation to be run on a single processor. In this case it is most efficient to spread the entire program over many processors. This *farm concept* for parallelization is illustrated in Fig. 6.1.

In general we have an algorithm which we label $\mathcal{A}$ and a set of parameters labelled $\mathcal{P} = \{p_1, ..., p_m\}$ for which we want to run the algorithm $\mathcal{A}$. For simplicity we shall assume that there are exactly $N = m$ processors available. If this is not the case then the set of parameters $\mathcal{P}$ can be divided into subsets appropriately. The *replication algorithm* is the assignment of the algorithm $\mathcal{A}$ to *all* processors $P_1, ..., P_N$ and the assignment of the set of parameters $\mathcal{P}$ to the set of processors N in a one-to-one (or many-to-one) mapping.

Since the entire set of processors has a homogeneous workload the speed-up is linear in the number of processors. Strictly speaking this is true only in the idealized situation where there is no loss due to the communication of results to the outside world.

We now consider dropping the requirement of a completely homogeneous task assignment to the processors. Assume that the overall algorithm $\mathcal{A}$ has k parts $\{\mathcal{A}_1, ..., \mathcal{A}_k\}$. Without loss of generality we assume that there is a set of parameters $\mathcal{P} = \{p\}$ and we anticipate that this set can be split up into m parts $P_0 = \{p_{0_1}, ..., p_{0_m}\}$.

The first case which we encounter is where all the partial algorithms $\mathcal{A}_i$ are identical, i.e., $\mathcal{A}_i = \mathcal{A}_j$ for all i, j. This is exactly the same as for the geometric parallel algorithms, which will be discussed in the next chapter, where each algorithm handles one part of a geometrically (i.e., homogeneously) decomposed problem. However, in general we have $\mathcal{A}_i \neq \mathcal{A}_j$ for all i, j and we need to replicate the partial algorithm to solve a specific problem. An example of this case is given below.

Let $C = \{I_1, ..., I_k\}$ be a partitioning of $I = \{1, ..., m\}$. The *partial replication algorithm* is the assignment of the algorithm to processors and the assignment of the set of parameters $\mathcal{P}$ to the set of processors p:

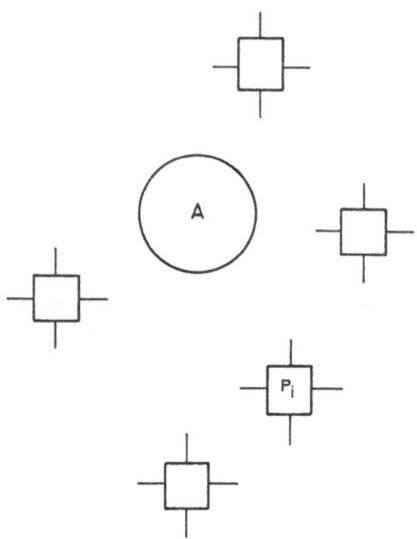

Fig. 6.1. Illustration of the farm concept for parallelization

$$\forall j \in \{1, ..., k\} \ \forall i_j \in I_j : p_{i_j}(\mathcal{A}_j, P_{0_j}) \quad . \tag{6.1}$$

The dropping of strict homogeneity implies that there must be cooperation and hence communication between the tasks. This is the fundamental difference to the homogeneous replication algorithm. In the full replication algorithm all tasks are independent of each other since they are operating on independent sets of parameters. In the heterogeneous replication the sets and the processes depend on each other. The communication between the processes is not of a trivial routing nature but involves manipulation of the supplied data. The speed-up is gained by the cooperation between the tasks and the level of partitioning of the problem.

Consider the multiplication of two matrices A and B. Let both matrices be $n \times n$ matrices. The two matrices can be multiplied by splitting them into smaller and smaller pieces until the pieces are so small that a single processor can handle one piece very efficiently. In such a scheme there are two types of partial algorithms. The first is the division into smaller and smaller parts and the joining of these smaller segments. The second algorithm is the actual multiplication.

In the first step we split the A matrix into two $n/2 \times n$ matrices as indicated in Fig. 6.2. The second matrix, B, is not changed. The first part of matrix A and the unchanged matrix B are sent to the next processor in a hierarchy and the second part of A together with the complete matrix B are sent to another processor. At this level the matrix B is split into two equally sized parts $n \times n/2$. The four pieces are then sent to the next stage. The parts of the matrices are now either further subdivided or are small enough for an efficient multiplication. After

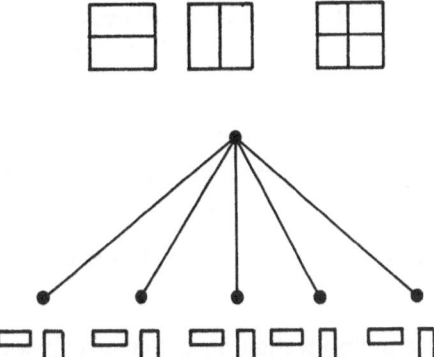

Fig. 6.2. Scheme of a partial replication algorithm for the multiplication of two matrices. There are two tasks to be performed: the division into smaller units and the multiplication

the multiplication of the partial matrices the results are sent up the tree. At each node the incoming partial matrices are joined up. At the apex of the tree the last two partial matrices are joined for the final result.

On the lowest level of this algorithm is the multiplication algorithm. Depending on the number of divisions there are k processes all carrying out the same task of multiplication. On the levels above the multiplication there are algorithms which partition the matrices, all performing the same task. In this example of a partial replication algorithm there are two types of algorithms running with different parameters, i.e., different parts of the matrices.

It is clear that the workload of different levels in the hierarchy of processors is different. The data flow is also uneven in the sense that the most data has to be transferred on the top level, while there is a decreasing amount of data flowing to the leaves of the tree where the multiplication is done. The processors are clearly not being homogeneously used, since while the multiplication is performed in the leaves no work is being done in the higher nodes.

Problems

6.1 Design a configuration of processors for the replication algorithm which has a better workload distribution than a linear list. Use only processors which have four links for communication.

6.2 Design a configuration of processors for the replication algorithm which has a better workload distribution than a linear list, but this time using processors which have m links (with $m < n$) to communicate. Is there an optimum communication structure for each pair (n, m)?

6.3 **Ray Tracing:** An interesting application of the replication method, although slightly modified, is ray tracing for computer graphics. Outline such an algorithm. (The reader is directed to the literature [6.1] for this application.)

6.4 Work out the transfer load for the multiplication algorithm given in the text. Can you invent a scheme which homogeneously uses all processors? Assume that all actions take the same amount of time.

6.5 Can you invent replication type algorithms for matrix multiplication which work for different processor topologies?

6.6 What is the complexity of the matrix multiplication algorithm which was given in the text?

6.7 Construct an algorithm for matrix multiplication on a matrix net of processors. What is the complexity? Can you show that there is an algorithm which is optimal? You are free to choose the processor topology.

6.8 The idea of recursive doubling can also be used for sorting a set of items. Construct an algorithm using such an idea. What is the complexity of your algorithm?

7. Geometrically Parallel Algorithms

A natural way to implement the simulation on a parallel computer of any system that involves a regular geometry and spatially limited interactions is to divide the volume into equally sized portions, each of which are then assigned to one of the parallel processing elements. This *geometric parallelism*, sometimes referred to as *domain decomposition*, can be implemented in a variety of ways, of which we present here some of the principle ones. The expression *data parallelism* is sometimes used in the literature to describe these types of algorithms, but we feel that the expression geometric parallelization describes more unambiguously the partitioning of the actual space in which the simulation takes place, and we reserve the use of the expression data parallelism to those cases where no such spatially oriented partitioning of the problem is implied, such as the algorithms described in Chap. 8. Although the methods we describe here are conceptually straightforward they require, in general, a substantial increase in program length and complexity. The additional time that is required to write and test such programs can be kept to a minimum by careful planning of the program structure and communication procedures between the processors in relation to the underlying geometry of the system being simulated. The reward for such effort is a program that uses the machine effectively and delivers as much of the available computing power as possible, thereby enabling us to study problems of a size that would otherwise have been inaccessible.

We will discuss some of the various possibilities for geometrically decomposing statistical mechanics systems in this chapter. In the introductory section we address some of the more general issues associated with such a spatially oriented partitioning. Local-update algorithms and cluster-type algorithms, which play an increasingly important role in reducing the problem of critical slowing down, are discussed in the context of spin models, and we look at questions of speed-up and efficiency in relation both to the particular system being simulated and to the computing and communication characteristics of the processors. The treatment of cluster algorithms introduces new and different problems for parallel simulations. A number of methods for the parallelization of molecular dynamics simulations are also introduced. The basic principles of these algorithms are outlined and we give a number of examples of how they may be implemented, together with problems that expand upon the concepts presented.

7.1 Geometric Parallelization

Before we proceed to introduce some methods for implementing geometrically parallelized algorithms it is necessary to recall and elaborate upon some of the concepts introduced in Chap. 4. There we discussed a number of different multiprocessor architectures, ranging from machines with a very large number of processing elements operating in a lock-step mode to machines consisting of a relatively small number of quasi-autonomous processors communicating over data links. Two of the key distinguishing features among the wide diversity of parallel computers are the degree of synchronization of the processing elements and whether the memory accessed by a processor is local or global (i.e., shared with other processors). The algorithms which we discuss are tailored to a set of independent processing elements with local memory which communicate over a message passing network. This is a very flexible type of computer architecture that we consider, and other more restrictive architectures such as lock-step parallel computers can be considered as being a particular case of this more general architecture. Consequently many of the algorithms which we discuss in this section can, with suitable modifications, be straightforwardly applied to other architectures.

The synchronization of processors via a message passing network is natural since the parallel algorithms which are used in computer simulations of physical processes are, by and large, *synchronized algorithms*. There are points in the algorithm (called interaction points) where processes need to interact and exchange data before they are able to proceed to the next step of the calculation. These interaction points effectively divide a process into distinct stages and, since the execution time of a particular process will in general be variable (depending upon its local data), the speed is governed by the slowest of the processes that have to be synchronized at a particular stage in the algorithm. For situations where synchronization is not necessary it is possible to overcome delays in message passing by using buffers appropriately.

In order to make the most effective use of the parallelism in a given situation we ideally require that each individual processing element is continually engaged in carrying out *useful* computation. This means that the time spent in communicating between processors should ideally be negligible with respect to the time spent doing computation, and also that the administrative overhead involved in distributing the simulation over a number of processors should likewise be negligible. In this situation we seek to achieve a good *load-balancing* of the individual processing elements, whereby each of the processors has the same amount of computation to do between the interaction points, and therefore no processor has to wait for another processor to complete its part of the computation before proceeding. The need to minimize the time spent in communication between processing elements is clear, since this represents time during which the computation idles, although some types of processors can carry out computation and communication independently, as we will discuss in Sect. 7.2.5. The question of the administrative overheads associated with the parallelization is rather more subtle, since there are a number of points that need consideration. It is clearly inefficient to duplicate some parts of the calculation on a number of processors, although it is often unavoidable. Likewise it appears inefficient to let any processor sit idle, but there may be efficient algorithms where this is actually necessary. This illustrates the point that such concepts as load balancing are not so much rules to be followed as guidelines to be considered, and every situation requires thinking afresh about the possibilities that it contains. A parallel algorithm will almost inevitably involve more steps (i.e., executed lines of code) than the equivalent scalar algorithm. It is important that this parallelization overhead, which depends upon both the type of algorithm and the degree of *granularity* [7.1] (i.e., the amount of computation to be done before communication is necessary), should not grow excessively as we increase the number of processors. Such considerations will become important when we consider non-local algorithms.

The method of parallelization that we address in this chapter of geometrically distributing the lattice over the available processing elements is an entirely natural one since the processor array then has the same geometry as the system being simulated. It clearly achieves an efficient load balancing if the system is reasonably homogeneous and has only short-range interactions and if each processor has the same fraction of the volume of the original system. There are a number of different ways of doing this and the efficiency in any particular circumstance will depend very much upon the physical model being simulated, the topology of the underlying lattice and the algorithm used for the simulation. It will also depend upon the details of the interprocessor communication, where it is important to consider whether there are dedicated links or a shared network that may involve possible interference and delays. These will all play a role in determining the effective data transfer rate between processors and the consequent efficiency of the algorithm.

As we mentioned earlier, the classification of the different parallel algorithms contains some arbitrariness and there is some overlap in the algorithms that fall within a particular category. It is sometimes useful to combine different sorts of parallelism within one algorithm to produce a *hybrid algorithm*. A simple example of such a hybrid algorithm is where each spatial domain of the system is assigned two processors, one which is responsible for the spin update algorithm and another which generates the required random numbers [7.2]. This scheme can be extended to situations where there are a set of processors, a *supernode*, associated with each spatial domain of the system. Each processor within this supernode may then have a different function, in exactly the same way as in algorithmic parallel algorithms [7.3]. Indeed, although we talk in this chapter of the system being geometrically distributed over processors, it is clearly possible to think instead of each processor being a supernode consisting of several processors with an internal communication and distribution of data.

The question of *dynamical load balancing*, i.e., of adjusting the spatial distribution of an inhomogeneous system over the processors to ensure that each processor has a roughly equal computational load, is one that has hardly begun to be addressed, but which will clearly play an important role in the future development of parallel algorithms. One straightforward way of achieving a partial dynamical load balancing is by the method of *scattered domain decomposition*, in which each processor is responsible for a number of widely scattered parts of the system. Consequently any fluctuations in the computing load in one part of the system will be compensated on average by having many such domains being calculated on one processor.

In this chapter we will first outline geometrically parallel algorithms for the Monte Carlo algorithm, using the two-dimensional Ising model as our example. The efficiency of this and other local-update algorithms is discussed. When sites are updated in parallel using the Metropolis algorithm it is necessary to take special care to ensure that detailed balance is maintained. The widely used checker-board pattern of updating to achieve this is described here. The standard communication procedures between processing elements, which do such tasks as collecting data from all the processors and sending it to the host-processor, are outlined here. The following section deals with geometric parallelism in higher dimensions (i.e., three and four) and the simulation of other spin and gauge models.

We then consider a non-local algorithm for simulating the Ising model, namely the Swendsen-Wang cluster algorithm [7.4], as discussed in Sect. 2.6.3. This algorithm, which is based upon the correspondence between the Ising model and percolation [7.5], is one of a number of cluster algorithms that substantially reduce the effect of critical slowing down for temperatures in the neighbourhood of the critical point, thus making it possible to generate independent spin configurations much more quickly. These algorithms require that we identify clusters of spins which are connected by bonds (that are distributed between neighbouring sites on the lattice with a temperature-dependent distribution as described in Sect. 2.6.3). The parallel implementation of such algorithms for identifying clusters, which are inherently non-local in nature, are described here in some detail. The method is illustrated using the Swendsen-Wang algorithm for simulating the two-dimensional Ising model on a many-transputer system [7.6,7] and the efficiency is discussed together with more general timing considerations.

Various algorithms for the parallel calculation of molecular dynamics systems are discussed, and the usefulness of the methods depends upon the range of the potential between the particles. In the case of long-range potentials in which the interactions between all pairs of particles must be calculated explicitly it is useful to implement an algorithmic scheme of parallelization [7.8,9], as described in Sect. 8.1. Some molecular dynamics systems involving solids can be parallelized essentially in the same way as spin systems. An algorithm for handling systems with relatively short-range interactions [7.10] is presented.

7.2 Strips, Squares and Checker-Boards

In this section we will discuss the basic techniques for distributing a lattice calculation that involves only local interactions over a number of processing elements. Each processing element is assigned responsibility for a distinct spatial subregion of the lattice. In order to carry out Monte Carlo updates of the spins it is necessary that the processors exchange information concerning the states of the spins on their boundaries. It is also necessary, of course, to combine the partial results produced by each processor for such global quantities as the energy and magnetization of the system. We consider first the spatial subdivision of a two-dimensional lattice and in Sect. 7.2.6 these methods will be straightforwardly generalized to systems in higher dimensions.

7.2.1 Detailed Balance and the Checker-Board

When we do an update of a single spin in the course of a Metropolis Monte Carlo calculation (Sect. 2.6.2) it is necessary to calculate the change in energy that would come about by flipping the spin. Once a decision has been made about accepting or rejecting the flipped spin we are free to go to one of the neighbouring spins and consider it for updating. However, if we were to try to independently update two neighbouring spins simultaneously then we would run into problems because the new trial energy of the bond between these spins is ill-defined and it is not possible to satisfy the condition of detailed balance (2.11,12). In a computation on a serial computer it is rarely necessary to give this point any special consideration, since it is in any case only possible to update a single spin at any time and therefore no such conflict can occur. On a parallel or vector computer, however, we need to be more careful in order to explicitly ensure that this situation does not occur.

The question that naturally arises is the number of sites on a lattice that can be updated in parallel. In a nearest-neighbour spin model, like the Ising model (2.53), the sites fall into two sets according to the checker-board pattern, as illustrated in Fig. 7.1. The sites are usually labelled "even" and "odd", according to whether the sum of the coordinates $(n_x + n_y)$ is even or odd. In the nearest-neighbour spin update using the Metropolis algorithm it is possible to update all the even (or all the odd) sites simultaneously without any conflict with detailed balance, but adjacent even and odd sites may not be updated simultaneously. This checker-board pattern represents the maximum possible parallelism (i.e., inherent parallelism), as discussed in Chap. 4 for the standard Metropolis Monte Carlo update procedure for spin models with nearest-neighbour couplings.

The details of how this checker-board pattern of updating is implemented will depend upon the particular computer. In some array computers (e.g., the Connection Machine [7.11] or the ICL DAP [7.12]) we may assign one spin variable to each processing element, in which case the Monte Carlo update pro-

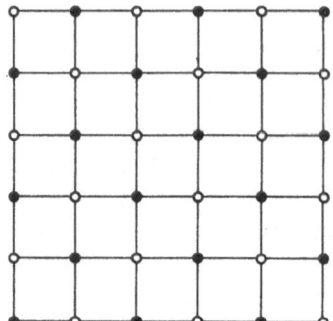

Fig. 7.1. The checker-board pattern of even and odd spins for updating

ceeds alternately on the even and odd sites of the array. The situation we shall discuss in more detail here is where each processing element contains a segment of the lattice with a substantial number of spin variables. In such a case the problem becomes one of ensuring that detailed balance is satisfied by the spins on the boundary of the lattice segment, i.e., when one or more of the neighbouring spins lie on another processing element.

This simple checker-board pattern only applies in the case of single spin updates for models with nearest-neighbour interactions. When we consider models with longer-range interactions it is necessary to build a new updating scheme that satisfies detailed balance. Alternatively, it is also occasionally useful to consider collective updates of groups of spins, in which case detailed balance requires special attention anew.

7.2.2 Strips

Assigning equally sized portions of the lattice to each processor is a natural way of ensuring that the system is load balanced, i.e., that each processor has the same computational load. For our example of the two-dimensional Ising model this is straightforwardly achieved on a P-processor system by dividing the L^2 lattice into strips of size $L \times (L/P)$, as shown in Fig. 7.2. The pth processor of a set of P processors is responsible for spins whose x-coordinate lies in the range $(p-1)L/P \leq x < pL/P$. The lattice sites on the interior of each strip may be updated using the Monte Carlo algorithm in the usual way, being careful to satisfy detailed balance (i.e., not updating two neighbouring sites simultaneously). The spins lying on the boundaries of the strips need special attention.

The processors themselves must then be connected in a *ring* structure where the neighbours in the forward and backward directions contain adjoining strips of the lattice. The ring structure provides a natural implementation of the periodic boundary conditions in this lattice direction, whereas the strip decomposition of the lattice means that each processor automatically accommodates the periodic boundary conditions in the direction contained within the strip. The interproces-

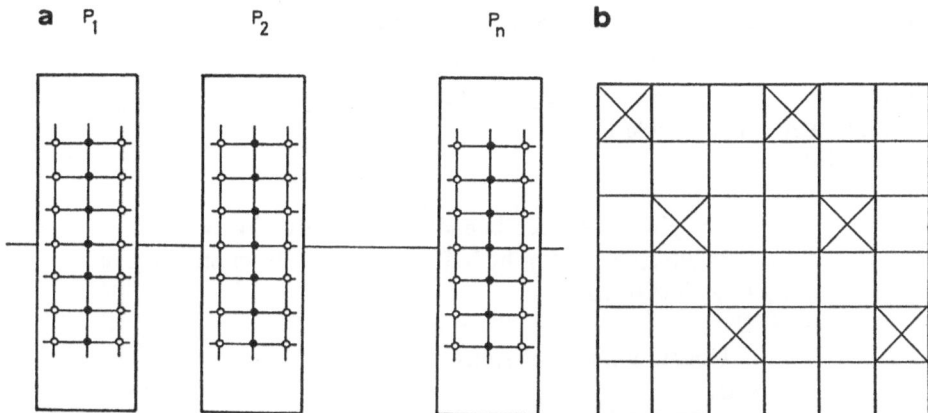

Fig. 7.2. (a) The geometrical decomposition of a two-dimensional lattice into strip sublattices, as used for parallel processors with local memory and message passing. (b) The virtual block structure within a strip

sor communication procedures for this processor topology are particularly simple, and will be discussed shortly. A simple alternative is to divide the lattice into squares of equal size, with each square being assigned to a separate processor. In this case it is necessary to communicate the boundary spins to the neighbouring processors in both directions, which also involves a two-dimensional *grid* communication network among the processors.

One simple way of ensuring that detailed balance is maintained while updating these boundary spins is to start with each processor sending copies of all spins along its two strip boundaries to the appropriate neighbouring processors. The Monte Carlo update then begins with the spins on the "upwards" boundary. As soon as these spins are updated a new copy is sent in the "upwards" direction cyclically for all the processors while they simultaneously continue updating spins in the interior of the strip. When the spins on the "downwards" boundary are updated this is done using the newly updated spins that have in the meantime been received from the neighbouring processor. These freshly updated spins are then sent cyclically downwards to complete one full update cycle of the lattice.

An outline of the most naive way of implementing this procedure is given for an arbitrary processor in the algorithm below, in which we use an Occam-like notation that is self-explanatory (for more details see Chap. 9). This algorithm closely follows the conventional sequential algorithm and it will provide an efficient implementation for processors that cannot carry out computation and communication simultaneously. However, for processors which can handle interprocessor communication independently of the numerical computation and without significant time loss it is more efficient to *tune* this algorithm so that computation and communication proceed simultaneously. An algorithm for achieving this is outlined in the following section.

Algorithm: Naive geometric parallelism in strips

Consider doing *mcsmax* Monte Carlo Steps on a lattice of size L^2 which is divided over P processors. Each processor is assigned L/P rows of spins. (We assume that L is divisible by P.)

    SEQ
      PAR
        ... Send spins on bottom boundary cyclically downwards
        ... Receive spins on top boundary cyclically from strip above
      SEQ *mcs*=0 FOR *mcsmax*
        SEQ
          ... Update top row of spins
          PAR
            ... Update spins on interior rows
            ... Send spins on top boundary cyclically upwards
            ... Receive spins on bottom boundary cyclically
          ... Update bottom row of spins
          PAR
            ... Send bottom row of spins cyclically
            ... Receive top row of spins cyclically

7.2.3 Squares

If the computing processors are capable of handling both computation and communication simultaneously then it is possible and desirable to optimize the above algorithm to make use of this ability. Such an algorithm proceeds by first subdividing each strip into P squares of size $(L/P) \times (L/P)$. The advantage of this procedure is that it is now possible to carry out the communication of the boundary spin variables and the spin-updating in parallel (which is not the case in the last step of the algorithm outlined above). This is done by ensuring that no neighbouring squares are updated in parallel and that the order in which squares are updated and their boundary spins exchanged across the strips is such that no square must *wait* for the communication about the state of the spins on a neighbouring square. There are, of course, many such possible patterns which ensure this.

Algorithm: Geometric parallelism in strips and squares

Our lattice is again of size L^2 and is divided over P processors, with each processor having L/P rows of spins (we assume that L is divisible by P). Each strip is now divided into P squares (or *blocks*), each of size $(L/P) \times (L/P)$.

    SEQ
      PAR
        ... Send spins on bottom boundary cyclically downwards
        ... Receive spins on top boundary cyclically from strip above

```
SEQ mcs = 0 FOR mcsmax
  SEQ
    SEQ block = 0 FOR P
      PAR
        ... Update spins within block
        ... Communicate the spins that lie on that part of the
            strip boundary between the next block to be updated.
```

The practical utility of this further subdivision of the strips into squares will depend very much upon the size of the lattice, the number of processing elements and the speed of the interprocessor communication. For many applications the naive algorithm is sufficient for all practical purposes.

7.2.4 Communication Procedures

One of the great advantages of geometrical parallelization is that the communication patterns that are required between the processors is typically very simple. In the examples that we consider here, the topology of the processors which naturally incorporates the periodic boundary conditions is a ring structure, whereby each worker-processor is connected to its neighbours in the forward and backward directions. This simple structure can be generalized to a grid structure in two or more dimensions by considering a series of rings. It is useful also to introduce a *controller* as a separate processor which is not part of the ring. The role of this controller is to exercise control over the processors in the ring, handle the communication with the host computer, accumulate the results from the processors in the ring and perhaps carry out part of the analysis of these results.

The basic communication patterns that are necessary are as follows:

- Input data from controller to worker ring:
  Parameters such as the desired temperature, number of Monte Carlo sweeps and so forth are broadcast to all the processors in the ring. It may also be necessary to input a set of seeds for the random number generator on each processor or an initial configuration of the degrees of freedom.

- Output data from the workers to the controller:
  Global quantities are cumulatively summed over all the processors – a process that has been descriptively called "harvesting" [7.3]. The final set of random seeds and the final configuration may also need to be output to the controller so that they can be used in subsequent simulations.

- Rotate data cyclically around the ring:
  This is typically the most intensively used communication procedure since it is necessary at each synchronization point to exchange data across the boundaries of the spatial domains.

This library of simple communication procedures forms the backbone of the communications structure for any geometrically parallelized system. Additional specialized communication procedures may be necessary in particular circumstances, such as those that allow any single processor in the ring to communicate with the controller or which allow any two non-adjacent workers to communicate, but these typically occur rather infrequently.

7.2.5 Timing and Efficiency Considerations

The efficiency of the geometric parallelism described here will depend principally upon the *perimeter effect*, namely the ratio of the amount of computation within a processor to the amount of communication that is required between the processors. The amount of data that must be passed between processors will depend upon the *perimeter* (or in three dimensions upon the surface) of the spatial area (or volume) of the system that is assigned to an individual processor. The amount of computation, on the other hand, is proportional to the area (or volume) associated with a single processor, and thus the perimeter-to-area ratio (or surface-to-volume ratio in three or more dimensions) provides a measure of the communication-to-computation ratio, and it is therefore a crucial parameter.

In this situation the expression for the efficiency of the algorithm on a multiprocessor machine (4.5) can be reexpressed in the form [7.9]

$$\eta_A = \frac{T_{\text{calc}}}{T_{\text{calc}} + T_{\text{comm}}} \quad , \tag{7.1}$$

where T_{calc} is the total calculation time and T_{comm} is the total time taken for control and communication.

Consider now the Monte Carlo simulation of the two-dimensional nearest-neighbour Ising model on an $L \times L$ lattice using P processors. We assume always that L is divisible by P and define $n = L/P$. We adopt the strip partitioning described in Sect. 7.2.2 in which the periodic boundary conditions in one of the directions are handled within the strip and $2L$ spins along the strip boundaries need to be communicated to neighbouring processors at each update step. The algorithm has been simulated on a transputer system with up to 128 transputers using various lattice sizes and numbers of processors [7.7], and the results are displayed in Fig. 7.3. A linear speed-up is represented by the straight line and we see that it is achieved for a small number of processors. But as the number of processors increases so do the communication overheads, and the efficiency eventually falls off, even for large lattice sizes.

In the case of transputer systems the hardware allows computation and communication to take place concurrently, and the above expression (7.1) must be modified by introducing an overlap time T_{overlap}. This is described and illustrated in [7.13] (see also [7.14,15]) for the geometrical parallelization of the solution of the Laplace equation, where the different effects due to non-overlapped and overlapped communications are illustrated.

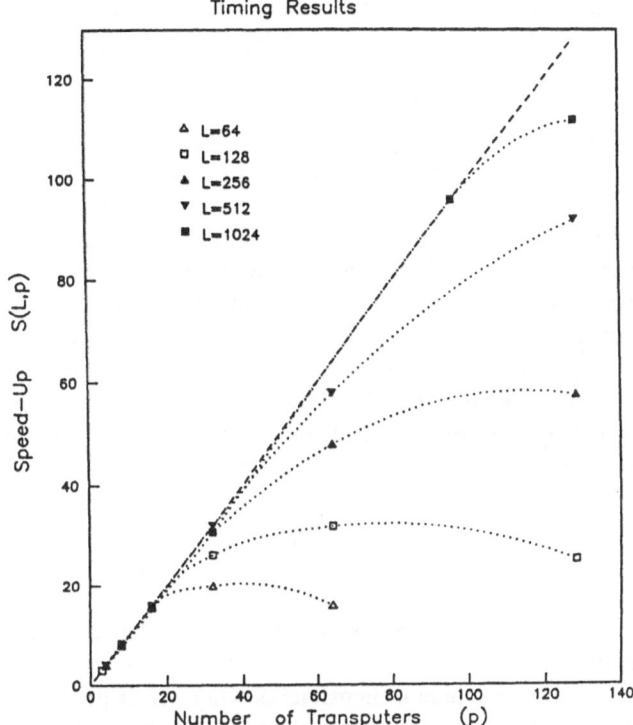

Fig. 7.3. Speed-up of the Metropolis algorithm for the two-dimensional Ising model with nearest-neighbour coupling

A comparison of the efficiency of geometric and algorithmic parallelism in the simulation of the X-Y model on a 16+1 transputer system has been made by *Askew* et al.[7.3]. Although the results depend, of course, upon the details of the processors, they indicate that hybrid algorithms involving both geometric and algorithmic parallelism are appropriate for a variety of problems.

7.2.6 Geometric Parallelism in Higher Dimensions

So far we have considered mainly two-dimensional systems, which have the advantages of involving only fairly simple processor topologies and which can be simply described and visualized. For systems in three dimensions it is possible either to generalize the structures we have already seen in two dimensions or to develop new three-dimensional communication topologies. We also look briefly here at systems in four dimensions, where there are a number of new and interesting possibilities for the processor communication structure. The interest in four-dimensional partitions comes from particle physics and lies in the simulation of lattice theories of the strong nuclear force, the relativistic quantum field theory called quantum chromodynamics (QCD). In such lattice gauge theory simulations the space-time continuum is approximated by a four-dimensional lattice of points between which a set of gauge link variables are defined. These link variables are typically 3×3 complex unitary matrices associated with the gauge group SU(3), and configurations of these variables can be generated using the Monte Carlo methods, as we have discussed for spin models. For more background and details of lattice gauge theory, the reader is referred to [7.16] and the references contained therein.

The simplest three-dimensional geometrical partition, and in many situations the most convenient and efficient, is the simple generalization of strips to *slabs*, in which the pth processor of a set of P processors is responsible for all spins whose x-coordinate lies in the range $(p - 1)L/P \leq x < pL/P$. The processor topology remains that of a ring and once again the periodic boundary conditions in the other two directions are automatically taken care of within the individual processors. This is also a partitioning that we will implement shortly when we consider parallel algorithms for molecular dynamics simulations. It is likewise straightforward to generalize the two-dimensional grid topology, in which each processor is responsible for spins whose x- and y-coordinates lie within a square, and in the three-dimensional case each processor is then responsible for a *square-stack* of spins. The interprocessor communication structure remains that of a two-dimensional grid, in which each processor is connected with its neighbours in only the x- and y-directions.

The next generalization of this in three dimensions is where each processor is assigned a regular cube, in which case the communication structure must also be three dimensional. The advantage that this gives in terms of homogeneity and good surface-to-volume ratio must be balanced against the increased complexity of the communication structure. For transputer-based systems, with their limited number of hardware communications links, it is useful to consider *supernodes* [7.17] or *cells* [7.14] of processors in which a small group of processors is assigned such a spatial subregion. Such cells or supernodes allow increased connectivity above that of the individual processors, as well as allowing either a spatial or algorithmic distribution of the algorithm among the processors of the cell, as discussed in [7.14].

A particularly elegant solution in three dimensions that is appropriate for some particular systems is to divide the volume into eight equal cubes. Each processor then only needs to communicate with three other processors, since the periodic boundary conditions ensure that the forwards and backwards neighbours in any one of the three directions in fact lie on the same neighbouring processor. This concept can be extended to partitions in four dimensions, where here the L^4 volume is divided into 2^4 *hypercubes* of size $(L/2)^4$ in which each hypercube processor needs only to exchange information with four nearest-neighbour processors. Although this is a very attractive solution to the problem of distributing a four-dimensional volume over a number of processors, it is limited in being restricted to a fixed number of processors, which may in turn limit the size of lattice that can be accommodated.

There are, of course, a vast number of other more flexible possibilities for the geometric partitioning of a four-dimensional lattice. One of the earliest such parallel simulations carried out [7.18] involved decomposing a $12^3 \times 16$ lattice onto a parallel computer by assigning each processor a $3^3 \times 16$ sublattice, i.e., carrying out a partitioning of just three of the four dimensions. *Hey* [7.13] outlines a number of other possibilities, such as the periodic square lattice, where only two of the directions are partitioned, and the repeated binary square, where the basic unit is a 2×2 square which is connected as a ring of four processors and can be repeated to yield an array of $N \times 2 \times 2$ processors onto which the four-dimensional lattice may be mapped.

7.3 Non-local and Cluster Algorithms

In Sect. 2.6 we discussed the advantages of using cluster algorithms to reduce the critical slowing down of simulations near a phase transition. These methods provide an enormous advantage over local update methods, but their disadvantage is the increased complexity of the program as compared with the standard local update Metropolis algorithm. This becomes particularly apparent on parallel computer architectures, which will be the subject of this section. The power of cluster algorithms, however, makes the additional programming effort both necessary and worthwhile. Work in this field is only at a very early stage and it is clear that there are many aspects to investigate.

7.3.1 Parallel Algorithms for Cluster Identification

The cluster identification method outlined in Sect. 2.6.4 is very convenient and fast on a scalar machine, but as we mentioned earlier there is at present no cluster algorithm that lends itself to vectorization. It is however possible to efficiently parallelize the algorithm, despite the fact that clusters may be spread over several processors.

We proceed at first in the same way as before by dividing the lattice into several strips, each of which is assigned to a particular processor, and the processors are connected in a ring topology. It is thus possible for two apparently different and widely separated cluster branches to be part of a single larger cluster, as was shown earlier in Fig. 2.1. This problem of cluster identification is clearly more severe now that the spins are distributed over several processors, since the branches of a single cluster may look like independent clusters within a processor and only be connected via a processor that is far away. We proceed in the first stage by treating each strip of the lattice separately so that the bonds along the two boundaries are considered to consist entirely of broken bonds and a "local cluster label" is assigned to each cluster in the strip. Clusters within the strip are built using a local stack vector in the way described previously in Sect. 2.6.4.

There are now a number of possible ways to connect these local clusters together into global clusters. The most naive way to solve this problem is to proceed by identifying all those clusters which lie on more than one processor (i.e., by looking to see if they are connected by a closed bond across any boundary) and then to iterate the standard cluster identification algorithm sequentially over all the clusters. The idea of the parallel algorithms is to implement this global cluster identification in an efficient way. In the following sections we present two such possible algorithms that acheive this goal and which we have called the *public stack* cluster algorithm and the *binary tree* cluster algorithm for parallel processors. In both these algorithms the first stage involves communicating the spins on the boundary with the neighbouring strip simultaneously with the generation of the actual bonds on the boundary, which we had considered to be broken in the initial stage of identifying the local clusters. The next step is to create at each boundary a "join vector" that contains the global cluster labels of cluster pairs that are connected across a boundary by a bond as well as the spin value. [The global cluster label on the pth processor in a series of P processors can be defined as $(p-1) \times (L/P) \times L$ plus the local cluster label, where each processor contains a strip of $(L/P) \times L$ spins, thus ensuring an overall unambiguous ordering of the clusters].

This way of organizing the problem of global cluster identification means that it is now only necessary to develop an algorithm that ensures that the join vectors are suitably used to connect all the local clusters. There are clearly a large number of possible algorithms, only two of which we present here in more detail. These methods can be tuned and there are any number of variants that are appropriate for particular processor capabilities, such as the *neighbour joining method* given in [7.7], but we have found the methods outlined here to be the most useful.

The non-local nature of the problem of cluster identification means that information must be communicated in some way between every pair of processors involved in the lattice simulation. This is unlike the usual (local) Monte Carlo algorithm described earlier, where it is only necessary for information to be exchanged between immediately adjacent lattice strips. This feature is the determining factor in the efficiency of the parallel algorithm, because the time spent in communication between processors increases not only with lattice size, but also with the number of processors over which the lattice is distributed. This is the source of the very different performance characteristics of the local and non-local algorithms, which will be discussed shortly.

7.3.2 The Public Stack Cluster Algorithm

The public stack cluster algorithm, which is described in considerable detail in [7.7], proceeds by collecting all the clusters that enter into the join vectors into a "public stack vector". Initially the public stack vector consists of an ordered list of clusters within the strip that are connected to the neighbouring strips. The join vectors and these initial stack vectors are communicated cyclically through all the processing elements. The stack vectors are accumulated and the join vectors are used to connect clusters in the public stack vector (in a similar fashion to that of connecting clusters in the local stack vector). After a full cycle of the join vectors through all the processors the new spin values on each processor can be read from the public stack vector (for clusters that were connected across a boundary) or the local stack vector (for spins in clusters lying entirely within a strip), and in this way a new spin configuration is generated.

Algorithm: Public stack algorithm for parallel cluster identification

---

Consider one spin update of an $L \times L$ lattice which is partitioned into strips over P processors, each of which is assigned a subregion of $L \times L/P$. The following steps can be carried out on all processors in parallel.

> SEQ
>> PAR
>>> ... Generate the interior bonds but keep boundary bonds explicitly broken
>>> ... Read/write boundary spins to/from neighbouring strip
>> ... Identify *local* clusters *within* the strip using local stack vector
>> ... Generate actual bonds on boundaries of strip
>> ... Send/receive array of *public* cluster numbers for all clusters on lower boundary of strip
>> ... Create *join* array of public cluster numbers where clusters are actually joined across the boundary by closed bonds
>> ... Create a *public stack* array that contains all clusters which are connected to clusters in other strips
>> ... Send/receive the join array and local section of the public stack array cyclically, and use them in each block to complete the public stack array as well as connecting the clusters
>> ... When the join arrays have done a complete cycle through all strips the public stack can be used to update the local stack array
>> ... The spins within each strip can now be updated from the local stack

---

Within the local and public stacks it is necessary always to use the rule that the cluster with the lowest cluster number is the seed. This ensures that the public seeds are correctly and unambiguously identified by each processor. After two start-up cycles, the cyclic send/receive can proceed simultaneously with the manipulations that need to be carried out on the public stack vector.

7.3.3 The Binary Tree Cluster Algorithm

In the public stack algorithm each processor essentially carries out a complete identification and joining of all clusters that are in the public stack vector. This is, of course, not strictly necessary since it is sufficient that one of the processors carries out this task and communicates the results to all the other processors. In the first stage of the binary tree cluster algorithm pairs of neighbouring strips are joined together and a *pair stack array* is created that contains all clusters which are connected across their common boundary. These pairs of strips are then themselves joined together in pairs using the same principle of connecting the clusters across the common boundary. In this way a binary tree is built by continuing this process of pairwise joining until the whole lattice is contained in the final pair. The information on cluster joining must then be sent down through the binary tree to the original strips, where the information is used to carry out the update of the spins.

7.3.4 Performance Measurements

As we discussed in Sect. 7.2.5 in relation to local-update algorithms, in the ideal case (i.e., where communication can be neglected) we expect to gain a linear speed-up as the number of processors is increased. The non-local nature of these algorithms means, however, that we very quickly run into a communication saturation situation as the number of processors is increased. The constraint of global connectivity that the algorithm imposes means that in practice the simulation is dominated by the communication when we have a large number of processors. At each stage of the algorithm it is necessary that each processor has the opportunity to communicate with every other processor, since only in this way can the global connectivity of the clusters be established.

The results presented in Fig. 7.4 on the timing of the public stack algorithm with various sizes of lattice were obtained by running the program on the 128 transputer machine of the Gesellschaft für Mathematik und Datenverarbeitung. The two-dimensional lattice was geometrically parallelized by partitioning it into P strips, where each strip was responsible for $(L/P) \times L$ data elements, as outlined in the previous sections. The machine, which is electronically configurable, was configured as a ring of processors. As in the case of the regular local-update Ising simulation in Fig. 7.3 there is a range of processors for which there is a linear speed-up. However, this range is very much narrower than for a local update algorithm, and as the number of processors is increased further the performance diminishes very rapidly and the simulation becomes communications-bound. This communication saturation sets in much earlier than for the local Ising algorithm due to the global communication that is necessary and also the rapid increase in the amount of data that has to be passed as the number of processors increases. It is clear from the performance results in Fig. 7.4 that it is essential to consider the optimum number of processors for a particular lattice size when using a cluster algorithm.

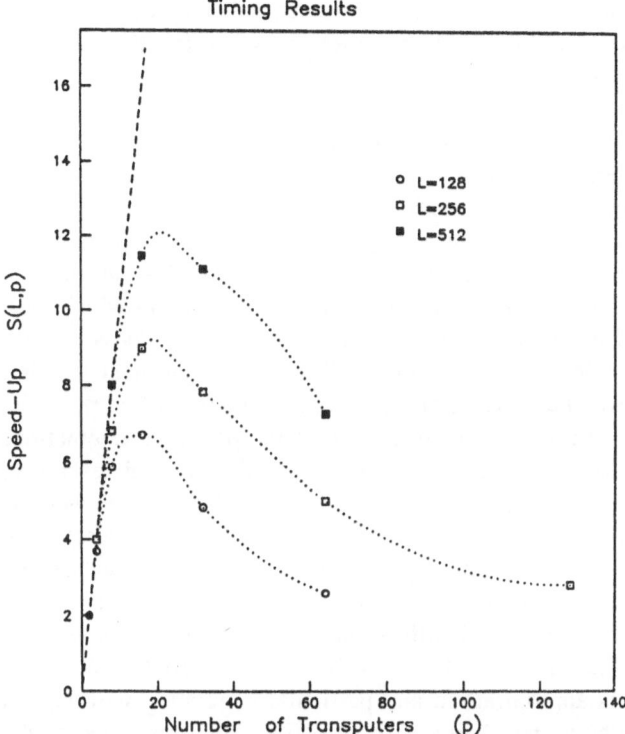

Fig. 7.4. Speed-up of the strip geometrically parallelized algorithm for the cluster problem

7.4 Parallel Molecular Dynamics Algorithms

The scale of computation involved in molecular dynamics simulations will be apparent from our earlier discussion. If we consider, for example, an N-particle system with long-range interactions then each step of the simulation involves the calculation of $N(N-1)/2$ interactions between pairs of molecules. For a moderately large number of particles N (i.e., $10^3 - 10^4$) followed over several thousand time steps it is clear that both very substantial computer resources and efficient algorithms are required. For such simulations there are clearly possibly very significant gains to be won in carrying out the simulation on a parallel computer. In order to actually achieve some substantial fraction of the theoretically available increase in computational power of these parallel machines it is necessary to pay particular attention to the range of the interparticle interactions.

7.4.1 Short-Range vs Long-Range Interactions

Consider first the case of a molecular dynamics simulation in which the range of the potential is very short – of the order of the diameter of the molecules. Each molecule then interacts only with a very small number of neighbours and the computing effort is of the order of the number of molecules N rather than N^2 as in the most general case of a long-range interaction. In such a simulation the major computational component involves identifying the interacting neighbours, while the actual calculation of the force between such interacting pairs of molecules requires relatively little effort. A parallel algorithm that is useful for such a situation [7.10] is outlined in the next section.

In the other extreme case of a long-range potential, such as a Coulombic or gravitational potential where it is necessary to calculate the contribution from all pairs of particles, a very different type of parallelization, in which each processor is responsible for a fixed subset of molecules (i.e., regardless of their spatial positions), is suitable [7.8,9] and this is discussed in the next chapter. However, even in the case where the potential is long range it is often possible, as discussed in Sect. 2.3, to restrict the calculation of the pair-wise interactions to nearby particles and to include the effect of more distant particles by an averaging process. Computationally this situation is very similar to that when the interactions are of intermediate range (i.e., over several particle diameters), such as occurs for fluids with a Lennard-Jones potential. In such intermediate cases the calculational load of a typical simulation consists essentially of two distinct processes, namely the search for and identification of interacting neighbours and the actual calculation of the interaction between the pairs of particles. Which of the above two strategies is appropriate in any particular simulation will depend both upon the details of the system being simulated (i.e., effective range of the interaction at the temperature being simulated) and upon the computer on which the calculation is carried out (i.e., interprocessor communication speed, memory per processor and so forth).

There are, however, situations where a geometric parallelization of the sort discussed earlier in relation to the Monte Carlo simulation of regular spin systems is appropriate [7.19,20]. This is the case when the system being studied is such that the molecules do not change their neighbours during the course of the simulation and we can regard the molecules as oscillating around some mean position on a regular lattice. It is then possible to geometrically parallelize the algorithm using the methods outlined earlier in this chapter for lattice systems. There is an added simplification for the molecular dynamics method over the Monte Carlo algorithm in that there is no constraint concerning detailed balance, since at any iteration of the molecular dynamics equations of motion the new variables depend only upon the values at the previous time step. This method has been used, for example, in studying the plastic-to-crystalline phase transition of sulphur hexafluoride [7.19,20]. The octahedrally shaped SH_6 molecules undergo a phase transition upon cooling from the so-called plastic crystalline phase to the true crystalline phase with a body-centred cubic lattice. During the course of the simulation the molecules can undergo orientational displacements which can lead to reorientations, and these studies have thrown some light upon our understanding of the plastic phase.

7.4.2 A Geometrically Parallelized Algorithm for Molecular Dynamics

In close analogy to the methods considered earlier for simulations of lattice spin systems on parallel computers it is possible also to construct a geometric parallelization of molecular dynamics algorithms for short-range interactions [7.10]. By assigning each processor responsibility for a spatial subregion of the system we can ensure that there is a good load balancing for the system. This algorithm is ideally suited to a network of P independent processing elements with local memory and a message passing communications scheme. As in our earlier discussion, the system is divided into stripes (called *slabs* here, since we will mostly be concerned with simulations in three spatial dimensions) and the pth processor is responsible for the molecules whose x-coordinates lie in the range $(p-1)L/P \leq x < pL/P$. The difference in the situation of molecular dynamics is that the molecules can move from one slab to another, and in doing so all their data is transferred to the new slab. The processors are conceptually organized as a ring and, as previously, the control processor, which distributes the data at the beginning of the simulation, supervises the collection of global measurements on the system (such as the energy) and so on, can be either a member of the ring or a separate processor.

The number of processors must correspond to the system size in the ways that we have outlined earlier. The amount of data transferred at each time step should be quite small in relation to the computation carried out, i.e., the granularity should not be too fine, and this requires a consideration of the slab thickness in relation to how short range the interaction is and the magnitude of the typical velocities. If the slabs are too thin then not only is there a large communication overhead, but it may also be possible that particles jump between non-adjacent slabs on consecutive time steps! We give below an outline of the algorithm that can be run simultaneously on each processing element. The most computationally intensive step of the algorithm, namely the calculation of the interactions, is carried out in parallel on all of the processors.

Algorithm: Geometrically parallelized molecular dynamics

Consider a complete time step of a molecular dynamics simulation in a volume of size L^3 which is distributed over P processors. Each processor is assigned a volume subregion (slab) of $L \times L \times L/P$ and the following steps can be carried out on all processors in parallel.

    SEQ
      PAR
         ... Send coordinates of molecules near slab boundaries to neighbours cyclically
         ... Receive coordinates of molecules near slab boundaries from neighbours
      SEQ
         ... Calculate interactions for all molecules (original and copied)

in slab

... Update coordinates and velocities of molecules belonging to slab (but not for the copied molecules)

PAR

... Send information concerning atoms that have left slab to the neighbours

... Receive information concerning atoms that have moved into the slab

---

When the molecules migrate from one processor to another we must ensure that the data arrays associated with the molecules do not continually grow during the course of the simulation. One useful means of accommodating this fluctuation in the number of molecules in a slab is the use of linked lists. By associating one linked list with the storage elements that are occupied and another with those that are vacant it is possible to avoid any need for data rearrangement. If there is very little movement of molecules from one processor to another then it may be sufficient simply to flag the array sites that are empty.

In this algorithm as it is given above the interactions associated with those molecules that are copied from one processor to another are computed twice. It is possible to modify the algorithm slightly to avoid this duplication of effort by copying the molecules in only one direction. The interactions that are required for these molecules are then calculated and passed back in the opposite direction to be combined with the partial results computed for the originals of these molecules. In most situations the number of molecules lying close to the boundaries of the slabs is small so that this modification does not have an appreciable effect upon the overall computing time.

The question of dynamical load balancing is potentially a very important one, but also one that is very difficult and has hardly begun to be seriously addressed. A situation where it could provide significant gains in efficiency is where a system is spatially inhomogeneous, such as occurs in condensation phenomena. If all the slabs in the above algorithm are the same size then it may happen that during the course of the simulation some slabs contain significantly more molecules than others. Consequently some processors will take considerably more time to complete their computation than others, thus giving a severe load imbalance. A dynamical load balancing would require some method of adjusting the slab thicknesses appropriately during the course of the simulation.

7.5 Hybrid Molecular Dynamics

Consider once again the N-particle system confined in a box, and for simplicity we consider here the two-dimensional case where the particles are confined to a square. In Sect. 2.4 we discussed the hybrid molecular dynamics method and indicated that it is possible to go from a completely global updating scheme to a more-and-more local updating scheme. This can be done by applying a geometric parallelization of the system to the hybrid molecular dynamics method [7.21].

Suppose we divide the square into smaller squares, as we have already done a number of times. This partitioning of the large square corresponds, in a sense that will shortly become apparent, to a lattice simulation similar to that which we are familiar with for the Ising spin system. In this analogy each of the squares can be considered as a type of spin variable which takes several values, and in the case we consider here the variables can vary continuously. The values represent the potential energy of the particles inside one of the smaller squares.

If we pursue the analogy with the Ising model further, each square interacts with the nearest and the next-nearest neighbours. It is now possible even to write down a Hamiltonian completely analogous to that of a lattice spin system. Recall that in the hybrid molecular dynamics method we compute a global move of all the particles by integrating the equations of motion forward. This positional change is either accepted or rejected, and in the latter case we try again with a new set of velocities. We now consider applying this scheme within a small square. We proceed by integrating the equations of motion for the particles in the smaller square as well as the nearest- and next-nearest-neighbour squares. Once the new positions of the particles inside the smaller squares have been calculated we can apply the acceptance procedure.

Detailed balance, however, requires some careful consideration. We cannot apply the integration scheme to all smaller squares at the same time. It is necessary to split the lattice into a checker-board-like decomposition and update the black and white squares alternately. This then ensures, together with the time reversal property of the integration scheme, that detailed balance is fulfilled. It is now clear that by dividing the large square into smaller and smaller squares we move from a global updating scheme to an increasingly local scheme, until we arrive back at updating single particles.

This procedure, which provides a continuous transition from a global updating scheme to a local one, enables greater freedom in adjusting the step size of the integration scheme. For very large systems consisting of thousands of particles it is necessary to choose relatively small step sizes in order to reach the optimal acceptance rate of roughly 50%. By partitioning the system into smaller units it is possible to choose larger step sizes since fewer particles are now involved in a single update step. The price to be paid for this advantage is that the number of particles involved in one updating step is no longer constant, and it is therefore more difficult to adjust the step size. With this scheme it is possible to update a system in parallel and at the same time use a non-local updating scheme.

7.6 Polymers on the Lattice

Polymer systems typically have extremely long relaxation times and it is necessary to generate a very large number of conformations of a chain in order that

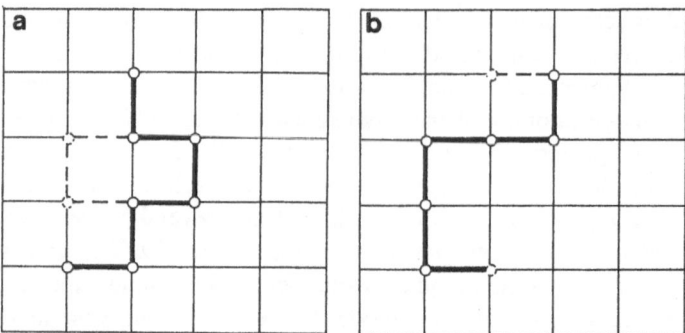

Fig. 7.5. A single polymer on a lattice, showing possible moves for a local Monte Carlo algorithm.
(a) The unit can move to a next-nearest-neighbour site or (b) a monomer unit can be removed
and added at the ends of the chain

a conformation is statistically independent of its starting conformation. Consequently the simulation of polymers requires computational algorithms that are highly optimized [7.22]. This is true not only for the static statistical information of chains but also when we are interested in the dynamics of a chain as it moves among other chains.

Even general types of polymer systems can be simulated using parallel algorithms and we are not restricted in this regard to polymer models that live on a lattice. In the next sections we will illustrate such parallel algorithms for Monte Carlo simulations of both a single chain and a dense system. We shall also look at off-lattice polymers as well as other parallelization ideas which involve some sort of geometric partitioning.

7.6.1 Single Polymers

In order for a single polymer to reach a new conformation which is statistically independent it is necessary to generate (at least in the worst case) of the order of $O(N^p)$ moves, where $p \geq 2$. Here N is the length of the polymer, which can be regarded as a string with N beads, as shown in Fig. 7.5.

We consider here Monte Carlo simulations of idealized polymer chains. Apart from a simple sampling of such a chain, where practically any move is allowed subject to a small set of restrictions (such as non-intersection and self-avoidance [7.22,23]), the principal sampling technique is that of importance sampling.

In the importance sampling method certain moves are proposed and either rejected or accepted with a given probability dependent on the temperature and the interaction between the monomer units. The move can involve, for example, the displacement of a monomer [7.23,24] or the displacement of larger units [7.25]. For our purposes here it is only necessary to consider one such possible move and we disregard the Metropolis proposition procedure by assuming that when it is possible to make a move (i.e., when no double occupancy of a single

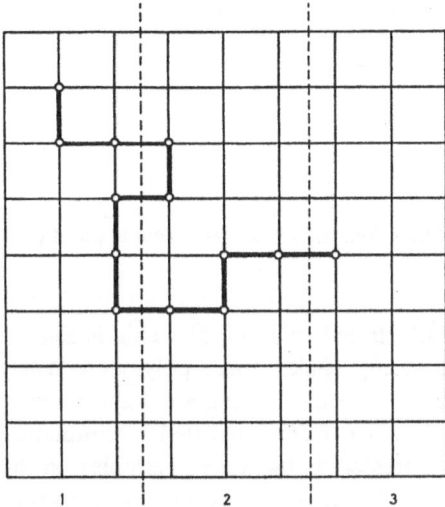

Fig. 7.6. A single polymer on a lattice where the lattice is geometrically divided into strips

lattice site occurs) then the move will be accepted. The move which we consider is indicated in Fig. 7.5b, in which a single monomer is affected. The lattice is taken to have periodic boundary conditions.

In order to parallelize the updating procedure of the single chain on the lattice we divide the lattice into strips, as indicated in Fig. 7.6. We first observe that, in general, the chain will not be stretched. Some parts of the lattice will be more densely populated and contain more units of the chain than other parts. This is simply a result of the moves allowed for a change of the conformation, in which a single monomer unit can move from one strip to a neighbouring strip. The workload will therefore not be as evenly distributed over the processors as was the case for the Ising model. Rather, the situation has some similarity with the geometric parallelization of off-lattice molecular dynamics.

7.6.2 Dense Polymer Systems

Consider now a dense system consisting of n lattice polymers of the type described in the previous section. The task of generating an entire configuration becomes quite formidable and it is necessary to drastically reduce the computational complexity in order to be able to simulate large systems. To achieve this reduction in the computational complexity we will use a geometric parallelization (or the dimensional reduction) of the system. This can be implemented by splitting the lattice into equal segments and by distributing them to the available processors in the machine. It is easy to see that for a sufficient number of processors, each operating independently on a segment of the lattice, the complexity is reduced to $O(N^{p-1})$.

We consider moves in which a single monomer is erased at one end of the

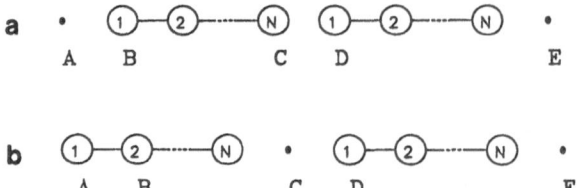

Fig. 7.7a,b The maximum "interference range" of the update of a void. (a) A void at site A pulls the polymer end at B, thereby hopping to site C (b)

chain and attached to the other end, as illustrated in Fig. 7.5b. This is known as the reptation method [7.22,26] or slithering snake. Each polymer consists entirely of monomers of either species A or of species B and a certain fraction, c_v, of the lattice sites are unoccupied. Since we were interested in simulating very dense systems ($c_v \ll 1$) the voids are treated as the active variables on the lattice. We proceed by randomly selecting a void and then choosing one of its four neighbours randomly. If the selected neighbour is occupied by the end of a chain, the void is moved to the other end of the chain, thereby "reptating" the chain into the former position of the void. Calculation of the resultant change in energy of such a trial move only requires data that is local to the two "active" lattice sites, namely the neighbouring sites of the old and new positions of the void. This is a direct consequence of the fact that we are dealing with homopolymers. The situation would, of course, be rather more awkward if we were to simulate, for example, heteropolymers or block co-polymers. In such cases we would have to explicitly calculate the energy changes along the length of the moving chain.

The fact that each trial move of a chain involves only local information renders a parallel implementation of such a simulation feasible. Any two chains which are sufficiently far apart may be updated independently of each other, and hence they can be done simultaneously. In this way it is possible to exploit the geometric parallelism in the model and preserve the processor-data locality. Different sections of the lattice may be allocated to a set of parallel processors, with neighbouring processors being responsible for neighbouring sections of the lattice. Each processor may independently attempt a trial move on some void in its own lattice section, as long as the distance between any two of these active voids is large enough to ensure that no two moving chains can interfere.

How large this distance should be depends on the current configurations of the reptated chains over which the voids may hop, which is of course a time-varying quantity. This is in contrast to the situation in models involving "immobile" lattice variables, such as in lattice spin models. In such cases the interaction range between lattice sites is time-independent and any two lattice points lying outside this range may be updated in parallel. We can nonetheless define a maximum "interference range" for the polymer model by considering the "worst-case" scenario shown in Fig. 7.7.

In Fig. 7.7 the void at A pulls the polymer end at B, thus hopping to the

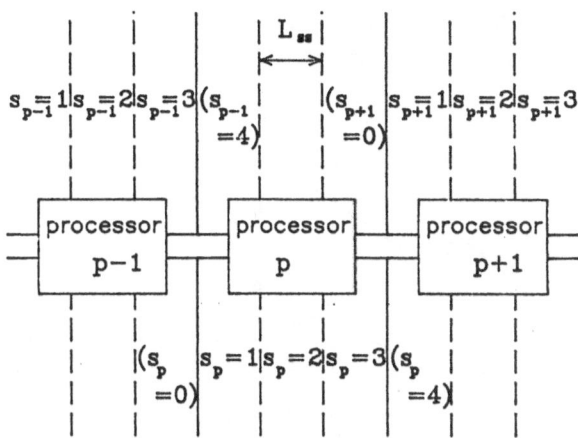

Fig. 7.8. A section of the lattice split up amongst three consecutive processors $p - 1, p, p + 1$ in the ring. Each processor i is directly responsible for (i.e., can move voids in) three substrips $s_i = 1, 2, 3$

other end C as shown in Fig. 7.7b. This move disables any other activity at the neighbours of C, and in particular at site D. If the end of another chain lies at D, as it does in this illustration, the polymer must remain fixed during the trial move of the void at A. Consequently a void adjacent to the other end of this polymer, at E for instance, cannot be updated. This, of course, holds equally for the other three directions from A. We see then that the largest possible range of interference of a trial move of a void is $2N + 1$ lattice constants. It follows that if two voids are to be updated in parallel they must be separated by at least $2N + 2$ lattice sites to ensure that no conflict can arise.

In view of these considerations, we can use the following geometric decomposition of the lattice amongst a set of parallel processors. An array of N processors are configured in a ring so that each processor is linked to its two neighbours cyclically. The lattice, L_D sites deep and L_W sites wide, is divided into N vertical strips of depth L_D and width L_W/N, so that each processor is allocated one strip. Each of these strips is further partitioned into three substrips of equal extent, each then being L_D sites deep and $L_{SS} \equiv L_W/3P$ sites wide. This subdivision is illustrated schematically in Fig. 7.8. Every processor, $p = 1, \ldots, N$, keeps three separate lists of voids, one for each set of voids residing in each of its three substrips, numbered $s_p = 1, 2, 3$ from left to right across the lattice strip.

With this partitioning of the lattice, N voids can be updated in parallel by demanding that every processor randomly selects a void from its substrip s, *for the same s* (i.e., $s_p = s$ for each p), except if we have $L_{SS} \geq N + 1$.

With such a geometric parallelization we indeed achieve an almost ideal speed-up. The speed-up curves for different settings of the lattice are shown in Fig. 7.9, in which a linear speed-up is indicated by the straight line.

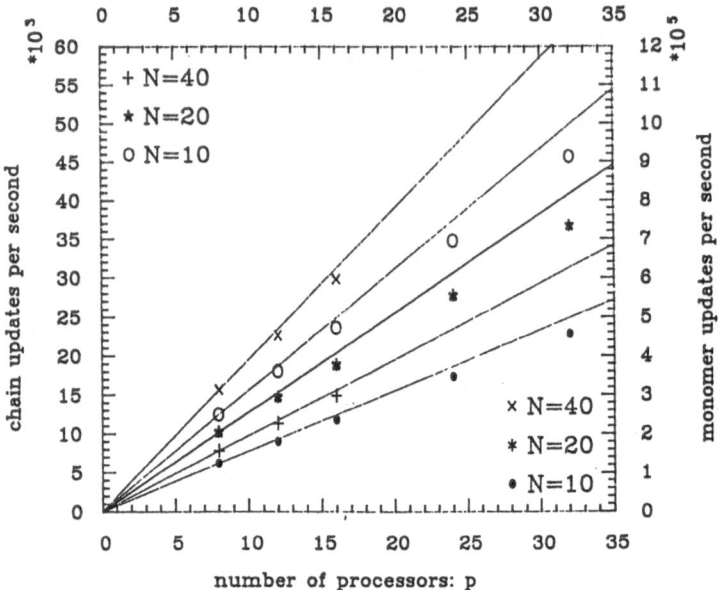

Fig. 7.9. The chain and monomer updates per second achieved on P processors. Three different chain lengths $N = 10, 20$ and 40 were used

7.7 Off-Lattice Polymers

In this section we discuss the simulation of polymer chains which are not on a lattice but in a continuum, and in particular we consider a single chain on a continuum two-dimensional surface. Every monomer unit can occupy, in principle, any position on the surface. Rather than considering the previous example of a single linear chain we instead address the question of studying branched polymers [7.22]. More specifically, we shall consider star shaped polymers, as shown in Fig. 7.10.

The central feature that is exploited in this case for the parallelization of the configurational changes is that the arms of the polymers cannot move arbitrarily within the plane, but are basically restricted to sectors [7.27]. It is therefore possible to reduce the computational complexity by splitting the two-dimensional space into sectors (i.e., the pie algorithm, see problems in Chap. 3, as is shown in Fig. 7.10b. Each sector contains at least one arm of the star polymer, and in general the number of arms-per-sector will depend upon the total number of sectors, number of arms and the stiffness of the chains. Such a decomposition means that only information from the neighouring sectors is required and it is possible to update sectors in parallel.

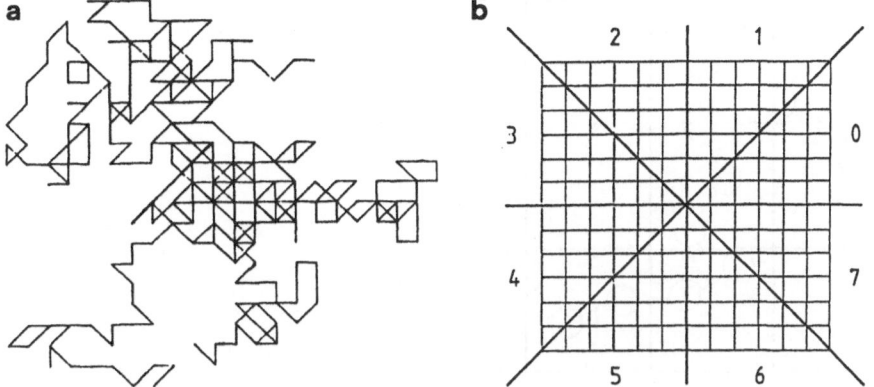

Fig. 7.10. A star-shaped chain polymer (a) and the geometric decomposition of space for the parallelization (b)

7.8 Hybrid Molecular Dynamics for Polymers

The hybrid molecular dynamics method [7.28] in conjunction with geometric parallelization can also be applied to polymer problems. The basic procedure as outlined above applies to the cases of both a single polymer and a polymer solution.

Figure 7.11 illustrates how a single polymer can be distributed over a number of processors by a geometric division of the underlying space. The density of squares depends upon the fineness of the grid into which the volume is partitioned. Such a partitioning is irrelevant for a regular molecular dynamics method, since every part of the volume is treated on an equal footing. However, using the hybrid molecular dynamics method it is possible to introduce larger changes in the conformations of the chain as well as being able to study the chain under constant temperature conditions.

7.9 Limits of Geometric Parallelization

As we would expect, there are limitations to the usefulness of employing geometric parallelization in the solution of various problems. From the speed-up curves for the Ising system, see Fig. 7.3, we have already seen the fall-off as the number of processors is increased. In the extreme case where the number of processors on which the simulation is carried out reaches the linear dimension of the lattice (i.e., when an $L \times L$ lattice is divided into L strips) the speed of the algorithm drops dramatically since the simulation becomes communication bound. This obviously represents a gross mismatch of the problem and the granularity of the processors.

However, there are more serious limitations to the geometric approach. In

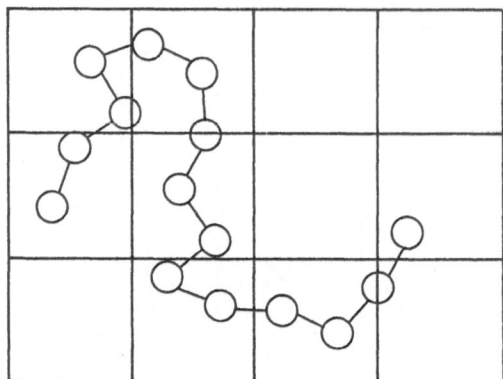

Fig. 7.11. A single polymer in a two-dimensional space and the geometric decomposition of space for the parallelization and the hybrid algorithm

order to illustrate this point we return to our familiar system of N particles in a box (2.24). Recall that in general the potential between two particles has a cutoff at a given distance r_c. Consider the simulation of a three-dimensional system at a density ϱ. The box length is then given by

$$L = (N/\varrho)^{1/d}   ,$$ (7.2)

where in this case the dimensionality d is three. Typical densities which one wants to consider in simulations lie between 0.2 and 0.8. With a system of 864 particles the linear dimensions so obtained lie between 16.28 and 10.26, whereas the typical cutoff for a Lennard-Jones potential is 2.5. Hence in a geometric parallelization into strips we can employ at most between 4 and 6 processors!

It would, of course, be possible to increase the number of processors to between 216 and 64 if we were to divide the cube into smaller cubes, but then we would run into problems with the connectivity. To illustrate these ideas we consider here only two-dimensional systems. The partitioning of these systems into strips and squares is illustrated in Fig. 7.12, together with the necessary flow of information. The ideal connectivity for such processors is 9 : 8, in order that they can both accommodate the interactions with the neighbouring squares and communicate the results to the outside world, although at present there is no processor or machine on the market which has such a connectivity. The required connectivity can be simulated by routing through other processors, and although such a procedure typically uses valuable processor cycles it may be possible to tolerate such a communication overhead.

A more serious consideration is the relation between the number of processors and the number of particles per processor. The largest number of MIMD processors that can at present realistically be connected together is of the order of a few hundred. To illustrate the problems involved suppose that we have 512 processors available and that we want to simulate a system with 40 000 particles at a density of 0.8. The average number of particles per processor is then

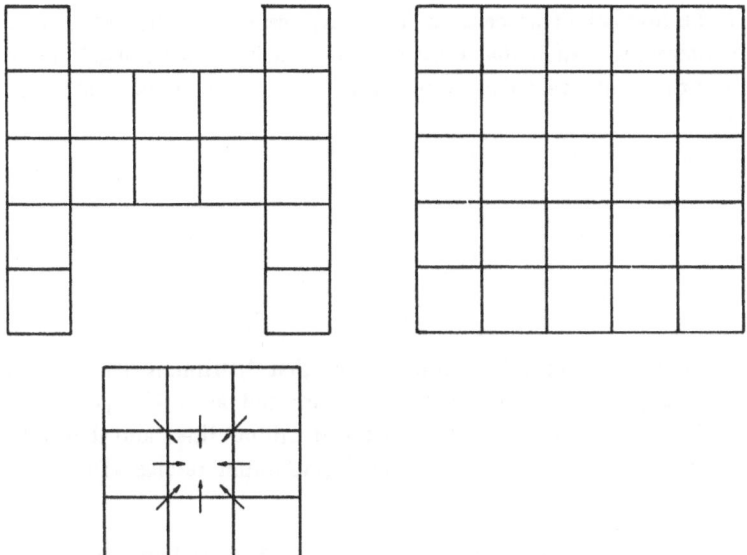

Fig. 7.12. Geometric parallelization for a molecular dynamics simulation. The flow of information necessary to update the positions and velocities within a single box is indicated

78, but because of the irregular distribution of the particles over the processors in a geometric parallelization it may happen that there are serious problems of load balancing in which some processors are almost idle while others have a heavy workload. Because the molecular dynamics algorithms are in general time-driven, it follows that the entire network of processors will be operating at a speed governed by the slowest. One must also address the question of whether the particle/processor ratio ($N = 40\,000$, $P = 512$) is in the regime where the speed of the algorithm behaves as $O(N)$.

These considerations highlight the necessity of having a network in which each of the processors is powerful. The problem of load balancing would become more serious if we were to employ a large number of medium-size processors. By reducing the number of processors the fluctuations in the number of particles per processor is diminished and a higher performance is achieved as a result of the better load balancing.

A physical situation where such a heterogeneous spatial distribution is exhibited is in a system that is quenched from a homogeneous one-phase state into the two-phase region. As time progresses the system develops regions which are much denser than the average density and eventually the system evolves into a state where only one such dense region remains. In the worst possible case this region is situated in only one processor, which now has to handle the entire workload, and the advantage of parallelizing the problem is totally lost!

The dense polymer system on the lattice also illustrates that there are restrictions to the number of processors that can effectively be employed on a given problem. The size of a single polymer chain determines the number of lattice strips and hence the number of processors. Typically the size of the chains that are of physical interest are such that only very few strips can be used. Thus the problem is not of a fine grain nature but rather demands a small number of powerful processors.

Problems

7.1 Write schematically a communication procedure that distributes data to, and collects data from, a set of 17 processors connected as a 2^4 hypercube. Assume that the processors have only four bi-directional links and that only one of the processors is connected to the input/output device (front-end machine).

7.2 Consider the binary-tree communication structure discussed in connection with cluster algorithms in Sect. 7.3.3. Write schematically a communication procedure for such a binary-tree.

7.3 In the geometric parallelization of a single chain on a lattice, what requirements does detailed balance place on the updating?

7.4 Making some reasonable assumptions about the average end-to-end distance of a single chain (e.g., a Gaussian distribution), what is the average time complexity for the geometrically parallelized (strip) single polymer chain problem?

7.5 For the simulation of a single polymer chain on a lattice, would a partitioning of the lattice into squares enhance the speed-up?

8. Data Parallel Algorithms

There are a number of situations where data parallelization is to be preferred over geometric parallelization. An example of such a situation that we have already encountered is that of a system with long-range interactions, such as Coulombic or gravitational interactions. In such a system all atoms or particles inside the computational cell interact with each other. A geometric partitioning would neither reduce the complexity nor spread the workload evenly among the processors, since all parts of the system must communicate with each other.

8.1 Data Parallel Algorithm for Long-Range Interactions

In the previous chapter we considered problems which could be effectively partitioned among several processors by assigning each processor a subregion of the volume. This decomposition of the problem is effective for most problems involving short-range interactions between the elementary constituents. However, in our discussion of molecular dynamics algorithms we saw that there are problems in which the interaction between every pair of constituents must be explicitly calculated, irrespective of their spatial separation. Consider two-body interactions of the following type, which are familiar from electrostatics and gravitation,

$$F_{ij} = G \frac{m_i m_j}{r_{ij}^2} \quad , \tag{8.1}$$

in which for a system of N particles there are $N(N-1)/2$ interactions to be calculated at any time step. Here there is clearly nothing to be gained by a geometric partitioning of the volume since this provides no simplification in the calculation of the interparticle forces.

An algorithm for implementing this problem effectively on a parallel computer has been proposed by *Fox* and *Otto* [8.1,2] in which each processor, of which we have say P, is randomly assigned a subset of the total number of particles N such that each processor has $n = N/P$ particles which it follows throughout the time evolution of the system. The spatial separation of the particles associated with any particular processor is unimportant. By assigning each processor an equal number of particles (which we call *local* particles) it is straightforward to achieve a good load balancing of the network. The processors are connected in a ring topology and the algorithm proceeds in the following way. Each processor starts by sending the mass and spatial coordinates of one of its local particles to the next processor in the forward direction around the ring, and at the same time receives a particle from its neighbour which lies in the other direction. The forces between this *travelling* particle and all the local particles are then computed. The travelling particle is then sent on to the next processor in the ring in a cyclical fashion and the whole procedure is repeated until the complete ring of processors has been visited. This entire procedure must then be carried out for every one of the n particles on each processor and the forces acting on each particle due to all the other particles are accumulated in order to move the particles forward one time step in the usual fashion.

This algorithm will provide an efficient means of parallelizing the problem as long as the time spent in communication of the particle masses and positions is small in comparison with the calculation of the forces. This can be arranged simply by ensuring that the granularity is not too fine, i.e., that there are a sufficient number n of local particles on each processor. In such a case this algorithm proves to be extremely effective.

A program written in Occam which carries out this algorithm written is given in Appendix C. In this program the host process, which distributes the initial parameters to the worker processes in the ring and collects the contributions to the total kinetic and potential energy from the particles in each process, is a separate process from the ring. Logically it could belong to the ring, and this can be achieved in Occam by placing two processes on the same physical processor, which would have virtually no effect upon overall timing since the computing load of the host process is minimal.

8.2 Polymers

The simulation of a single polymer chain on a lattice using a geometric parallelization of the lattice into strips or squares suffers from the problem of an uneven workload among the processors which operate on the parts of the lattice. Each segment of lattice (in this case strips) contains, in general, an unequal number of monomer units. This problem can be overcome by using a data parallel algorithm where the number of monomer units per processor is fixed.

In the data parallel algorithm we divide the polymer chain into equal units of groups of monomers. This is shown schematically in Fig. 8.1. Each processor is then responsible for N/P of the N monomer units, where "monomer units" are to be understood in a very broad sense. In some situations it may be better to use a different partitioning, for example into the repeat units of the polymer. In both cases there will be a fixed number of units per processor. For such a partitioning it is not relevant that the units are sitting on a lattice and we do not distinguish between lattice and off-lattice types of simulations.

The equal division of the data into parcels has the consequence that some of

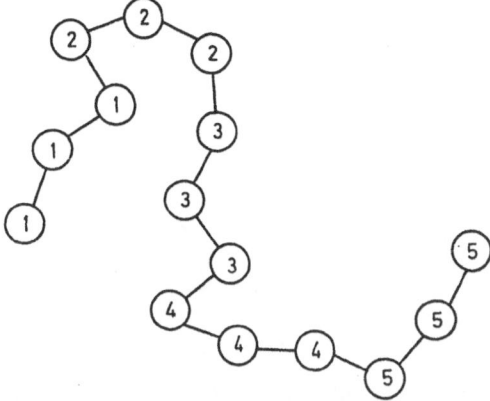

Fig. 8.1. A single polymer in data parallel form. The numbers indicate the mapping to the processors

the neighbour relations are preserved here in the polymer problem, whereas for monatomic systems the neighbour relations are completely lost in a data parallel algorithm. Atoms which live on the same processor may be so far apart that they do not interact. In the polymer problem the preservation of part of the neighbour relations yields a reduction in the amount of computation and communication necessary to generate the next conformation of the entire polymer chain.

The most straightforward configuration of the processors to handle the communication is the ring topology. Since there are possible connections between all data parcels, every processor needs to get data from all the other processors. However, there are some polymer problems where the conformation is fairly rigid and this feature can be used to reduce the communication overhead, although this is not generally the case for monatomic systems. In such situations it would be desirable to have an adaptable topology for the processor connections, something which in principle is possible with transputer-based machines. If the conformation is slowly changing, or even fixed, and one only wants to compute some static properties of the system, the processor topology can be adapted to the neighbour relations of the particles and the processor topology would follow the conformation of the chain. This would result in a considerable reduction of the computational complexity.

So far we have only considered the simplest case of a linear polymer chain. A different case of considerable interest is the situation where the chain has side arms, as indicated in Fig. 8.2. Several types of partitionings are possible for such a case. If the polymer has a repeat unit where each unit comprises a side arm and some monomer units of the backbone, it is possible to use this repeat structure of the chain. Each processor then handles a segment that includes part of the backbone and a side chain. The second possibility is to group together a number of monomers, irrespective of whether they belong to the backbone or the side chains.

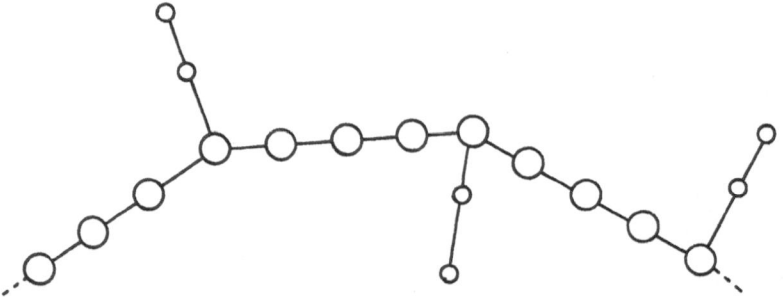

Fig. 8.2. A single polymer chain with side arms

Both partitionings make good sense. Although the first type of partitioning is usually the most appropriate, there are situations where the polymer may have only a small number of side arms in comparison with the number of monomer units in the entire chain and where this type of partitioning would therefore not result in a reduction of the complexity of the computation. In this case the second algorithm would be much more appropriate. The type of partitioning depends upon the granularity of the processor. If there are only a small number of very powerful processors then the first procedure is more appropriate.

'The situation for star-like polymers is more intricate. Again one could split the polymer into monomer units irrespective of their relation to the arms. However, there is a second possibility, which involves splitting the polymer hierarchically. In a case such as that of star-like polymers not all the arms can interact with each other, and hence a hierarchy reduces the number of interactions between the arms.

Problems

8.1 Consider the Hamiltonian $\mathcal{H}$ [8.3] of a polymer chain consisting of the following three parts:

$$\mathcal{H}_j = \frac{1}{2}k_b(l_j - l_0)^2 \quad ,$$

where k_b is the spring constant,

$$\mathcal{H}_{\theta j} = \frac{1}{2}k_\theta(\cos\theta_j - \cos\theta_0)^2$$

for the valence angle (θ_0 is fixed) and

$$\mathcal{H}_{\Phi j} = \frac{1}{2} k_\Phi \sum_{n=0}^{5} a_n \cos^n \Phi_j$$

for the torsional angles. In addition to these interactions there is also a Lennard-Jones 6–12 potential for the interactions along the chain backbone. Calculate the complexity of the problem using a data parallel algorithm if there are as many processors as there are monomer units.

8.2 What alternatives are there to the ring processor topology for the linear chain polymer problem?

9. Introduction to a Parallel Language

So far we have presented parallelization methods, strategies and parallel algorithms for computational science problems. What we would like to untertake now is to give an introduction to a programming language with which it is possible to realize the parallel algorithms on parallel machines.

9.1 Transputer-Based Parallel Machines

From a computational science point of view the transputer offers an enormous amount of freedom. With the transputer it is possible to build machines with an almost arbitrary topology. There is no rigid communication net or synchronization.

The transputer [9.1–4] is a processor with integrated communication facilities. It is a building block which one can bolt together to make large multiprocessor architectures (Fig. 9.1) tailored to the algorithmic needs. This may involve a topology closely following either the data structure or the flow of data in the algorithm.

The main feature is the four communication links which are integrated into the processor chip. One can couple transputers via such links and the communication between transputers through these links is bi-directional. Hence eight simultaneous messages can be passed between the transputers. There is no common bus via which messages or data items can be exchanged. In a sense this is similar to connecting a large number of computers via their serial ports.

Later on, when we discuss the Occam language, it will be of use to us to know how the eight transputer links are labelled. The links are labelled as shown in Fig. 9.2. The four out-going links have the labels 0–3 and the incoming ones 4–7. In the 20 MHz version a link is capable of transmitting at a peak rate of 20 Mbits/s. The measured values [9.5] are per link unidirectional 1.82 Mbytes/s and bidirectional 1.52 Mbytes/s.

The communication between transputers can be done simultaneously with computation in the arithmetic unit. In the T800 version of the transputer the arithmetic unit is integrated into the chip. The unit is a floating point unit which internally works with 64 bits but the data path to the unit is only 32 bits wide. The performance of this unit was measured [9.5] to be 0.77 MFlops for a multiplication and 1.25 MFlops for an addition. A speed of 1.74 MFlops was measured for two consecutive adds and 1.2 MFlops for an addition followed by a

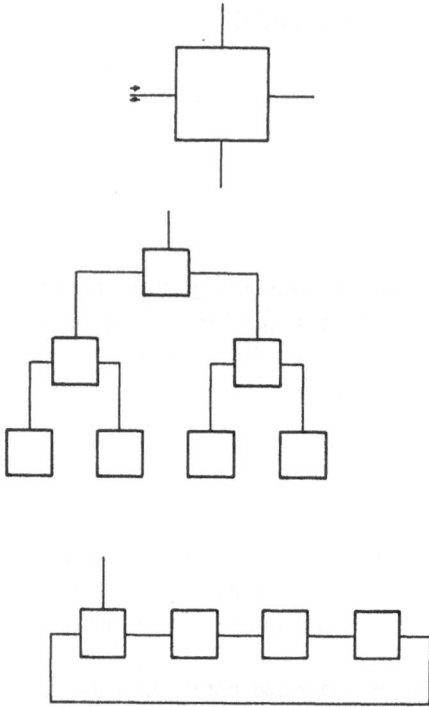

Fig. 9.1. A schematic representation of a transputer with its four links (*top*). With such a building block a variety of topologies accommodating many algorithmic needs can be realized (*centre* and *bottom*)

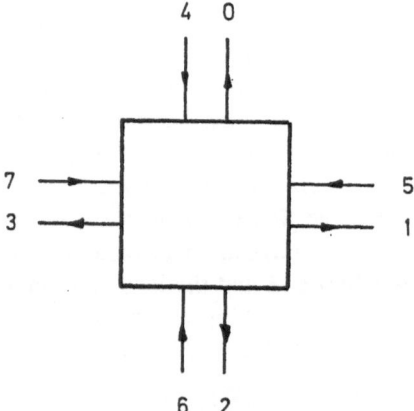

Fig. 9.2. Numbering of the four incoming links and the four outgoing links

multiplication. Using integer arithmetic one can achieve a speed of 2.44 MIPS (million instructions per second), while the integer multiply runs at 0.43MIPS. The wide range of these numbers show that a code with mixed operations has to be carefully tuned in order to maximize the speed.

In addition to the integrated floating point unit the processor features a 4k fast internal RAM. This memory cannot be directly accessed from other transputers. A transputer-based machine hence falls into the category of a local memory machine since all access to memory is controlled by the transputer to which the memory is connected. The functional units of the processor are shown in Fig. 9.3. The transputer can, of course, be connected to external RAM.

A single transputer can handle more than one process, and when this is the case, a scheduling between them is carried out. This is important for some applications, particularly if there is simultaneously a communication and a compute process. Moreover, it simplifies the development of a parallel program, since it can be designed and tested on a single processor. Once the validity of the program has been established, the different parts of the program can be distributed to other transputer processors for true parallel execution.

As well as the basic building block one also needs a connectivity structure for large machines. Most transputer-based machines have different possible conectivity between the transputers. The Parsytec machine [9.6] follows a hierarchical concept, which allows us to connect 16 transputers whichever way is desired. On the next higher level four such computing clusters of 16 transputers can likewise be connected in any way one wishes, as illustrated in Fig. 9.4. Unfortunately, other manufactures do not disclose their machine connectivity.

The Meiko Computing surface [9.7] and the Parsytec Supercluster are based on the transputer and are electronically configurable, i.e, the user program determines the topology of machine within the limits of the possible graphs the connectivity concept can handle. If the connectivity is not sufficient to embed a certain graph, routing processes have to be inserted. One processor is then occupied by more than one process.

9.2 Parallel Programming in Occam

So far we have discussed parallelism from an abstract point of view. Here we want to give an introduction to a programming language which allows us to formulate parallel algorithms. Our emphasis is on displaying some of the concepts behind the parallel programming language, and the examples given focus on applications in computational science.

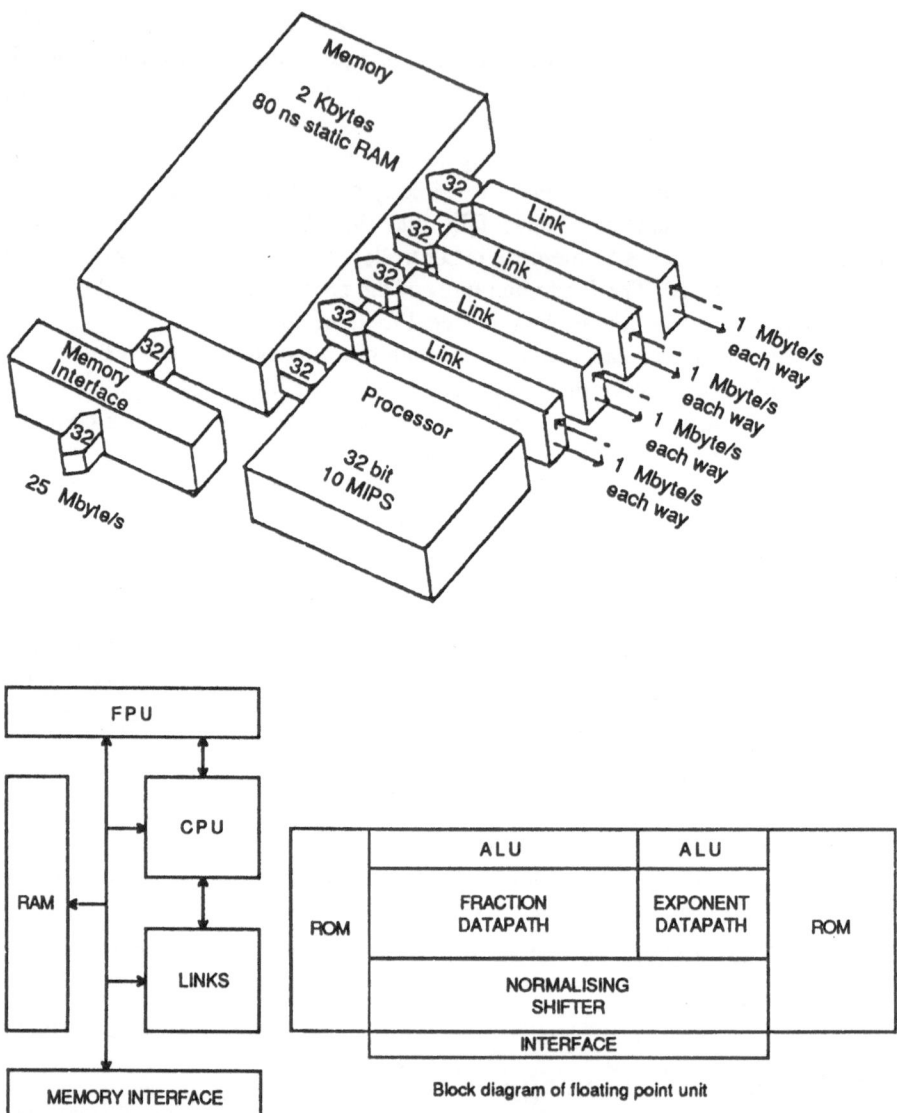

Fig. 9.3. The functional units of the T800 series transputer

The exposition of the language will be very free and we do not necessarily cover the material in a "logical" progression. Since the reader of this book is expected to have been exposed to other programming languages we can assume that he/she is familiar with the basics of such languages. Hence we may use language elements at points before they are actually defined or discussed whenever we believe that the meaning is clear from the context.

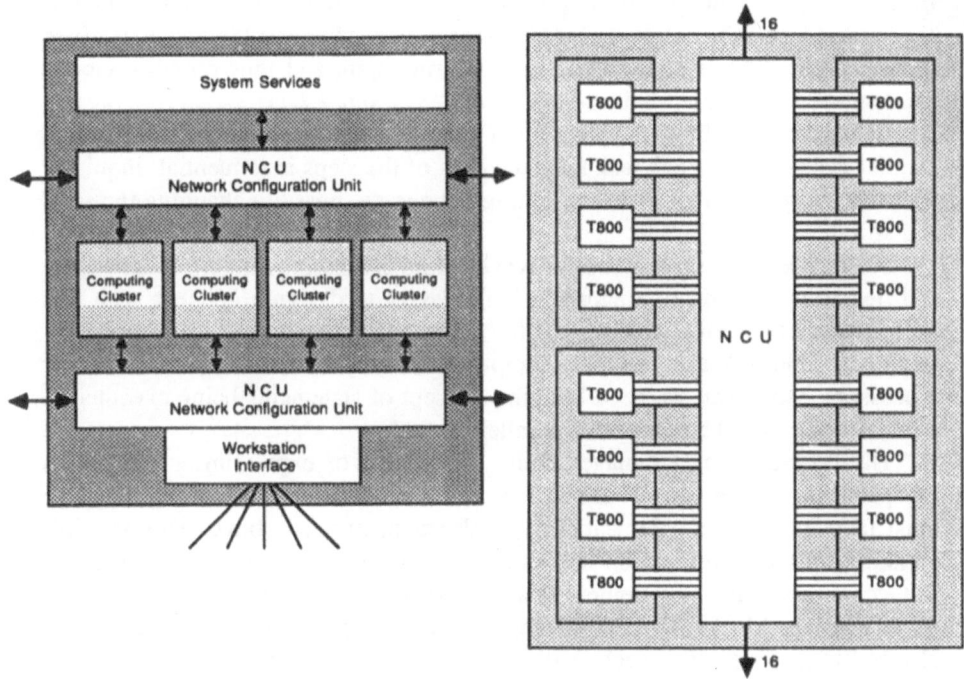

Fig. 9.4. Design details of the Supercluster transputer-based parallel machine. On the lowest hierarchical level are the computing clusters

This is not meant to take the place of a manual or a reference book. We assume that the reader will, either subsequently or while reading this chapter, consult other texts dealing explicitly with the Occam language. For this purpose we have compiled a list of books dealing exclusively with Occam [9.8–11]. This list is far from complete since it represents a choice that is biased towards our goals with respect to computational science, and we apologize to those authors who are not listed.

9.2.1 The Process and the Channel Concepts

Consider a simulation of an Ising spin system in which a processor is assigned to every spin on the lattice. Each spin in this system can be handled individually [9.12,13] such that detailed balance for the Monte Carlo procedure is fulfilled as long as we know exactly all the states of the neighbouring spins at any given time. Here we run into a problem because each spin (i.e., each processor) is working on its own but from time to time needs to know the condition of the neighbouring spins or processors in order to change its own state. This change of the spin state must be screened from the other spins, i.e., it should not interfere with other spins although all the processors run in parallel. If the spin change is not screened then the neighbouring processors do not know if they have received an old state or a new state of the neighbouring spin, and thus possibly violate the condition of detailed balance (see Sects.2.2.2 and 7.2.1).

Hence we can split up the problem of generating a new spin configuration into parallel tasks. Within each task the execution of the steps is sequential: inquiring about the neighbouring states, computing the energy, possibly changing the spin, and so on.

More generally we can consider two or more processors running concurrently and operating on a data structure. It is necessary to design a concept such that no processor interferes with any of the others. In other words, each processor should be able to make certain assumptions on the condition or state of other processors. Moreover, we need both the concept of statements being executed in a linear order and the concept of parallel execution.

The programming language should be capable of expressing an algorithm by specifying which parts could be run concurrently. Furthermore, the language must take into account the screening problem mentioned above. This idea off breaking up an algorithm can be followed right through to the statement level. In Occam every line or statement is a *process*. Processes can be executed in the same sense as blocks of code in other programming languages. Processes can also be single statements, such as the addition of two numbers. The processes or parts of the algorithm can, at least in principle, run concurrently.

Consider what would happen if we were to arbitrarily split a program into a number of separate parts and run these concurrently. Within each part of the program the execution of the statements would remain linear, i.e., in the order that they appear. However, the outcome of the execution of the entire program would in general be unpredictable, since there is no order among the parts, and consequently no way of predicting when a variable would be accessed. To be able to make certain assumptions on the state of the parts we need to communicate or *synchronize* at certain points in the overall program. At these synchronization points the state of the variables in different parts of the program should be predictable. The way that Occam implements this idea is by the *channel* concept.

To illustrate the above points we consider again the Ising model and, for simplicity, we take a one-dimensional chain, as illustrated in Fig. 9.5. We want each spin to be handled by a process which does the following:

• Inquires about the states of neighbouring spins

• Computes the probability of changing the state of the spin

• Possibly changes the spin state

All of the N processes, which we call *spin.process*, should run concurrently. This is expressed by the following **PAR** construct:

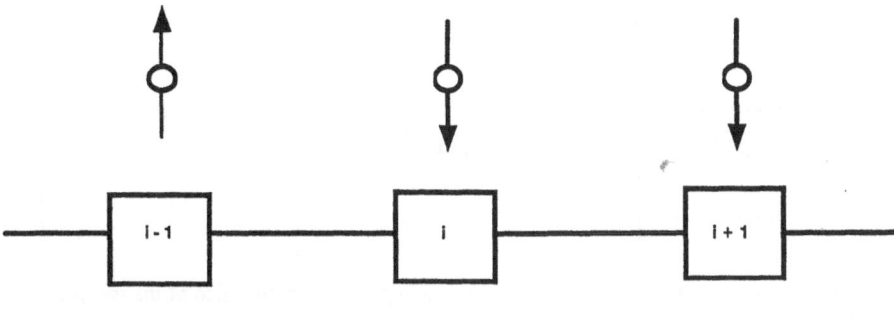

spin process

Fig. 9.5. One-dimensional spin chain and the corresponding processes

PAR
    spin.process.0
    ⋮
    spin.process.(N-1)

Within each of the N processes the execution of the statements should be done in a sequential order. This is expressed by the **SEQ** construct:

SEQ
    Inquire about the states of neighbouring spins
    Compute the probability of changing the state of the spin
    Possibly change the spin state

More formally a process can be a single action, such as the addition of two numbers, and such a process is called a primitive process. Assignments are also primitive processes. A process can have more than a single action, in which case it is called a compound process. A process starts, performs these actions, and then terminates. In our example we have a process in which the action is the parallel execution of N other processes, each of which also performs some further actions.

When does a process with parallel processes terminate? A sequential process terminates once the last statement has been executed. A parallel process terminates once all the constituent processes have terminated.

So far we have not solved the problem of screening. The inquiry about the state of a spin cannot be done by reference to a global data structure since the data structure is local to the process. In such a case where the data structure is local we must have a means by which the other processes can inquire about the state of a spin. The concept that Occam uses is that of *communicating sequential processes* [9.14] in which a message is sent from one process to another.

These messages are passed via channels (Fig. 9.6). and the channels have names by which they are referenced within a process. If two processes have agreed to communicate via the channel with name *message.channel* then messages can be passed between them. In our example, each process needs four

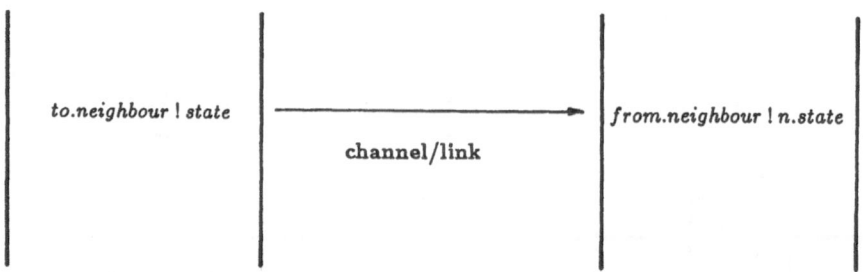

Fig. 9.6. Illustration of the channel concept and the implied synchronization of the two processes involved

channels, two for each of the neighbours for reading and writing. Over these channels the *spin.process* receives the states of the neighbours and sends its own spin state.

PAR
$spin.process.0(chan2_0,chan1_0,chan1_1,chan2_1)$
$spin.process.1(chan2_1,chan1_1,chan1_2,chan2_2)$
$\vdots$
$spin.process.N\text{-}1(chan2_{N-1},chan1_{N-1},chan1_0,chan2_0)$

Channels are much like a generalized storage since we can put a data item into a channel or read it from a channel. To make this connection even more explicit, a data item is either written or read from storage or from a channel as follows:

- $a := b$

- $a \ ! \ b$

- $a \ ? \ b$

The first is the usual assignment of the value of the variable b to the storage element with the label a. This constitutes a primitive process. The second and the third statements represent the sending and receiving of messages via channels. The value of b is sent via the channel with the label a by using the exclamation mark. If a message arrives via the channel a, then the question mark means that the message will be assigned the value of b. Both of these are primitive processes and it should be noted that only a *value* is sent or received.

In our example of a spin update we can now write the appropriate code for the receiving of the state of the neighbouring spins and the broadcasting of the local spin state to the neighbours. The receiving and the sending of the spin states can, of course, be done in parallel, so that we can write

```
SEQ
  PAR
    left.neighbour ? state.left
    right.neighbour ? state.right
    broadcast.left ! own.state
    broadcast.right ! own.state
  Compute probability of changing the spin state
  Possibly change the spin state
```

Note that after the keywords **SEQ** and **PAR** we have indented all subsequent statements. The indent is a syntactical unit much like the {} brackets in C or the **begin** and **end** in other languages. The indent is always by two spaces and as soon as the block is finished one has to outdent again to the level at which one wants to continue.

Note that the above spin update algorithm is not yet complete since we must take detailed balance and some other points into consideration, as we will explain shortly.

Sending messages via channels implies that there is some sort of synchronization mechanism. In the above example the synchronization is indeed very tight since the N processes perform message passing and communication of spin states in unison. No process can proceed or advance ahead of the others until the message passing is complete.

In Occam the message passing is synchronous. Two processes which want to exchange data perform a *rendez-vous*, exchange the items and then decouple from each other. There is no asynchronous message passing between processes, i.e., there is no implicit buffering.

In our discussion here of the process concept we have not said anything as yet about how the processes may physically be distributed over a set of processors. The language elements that enable us to assign processes to physical processors will be described in Sect. 9.2.9. One of the particularly nice features of Occam is that this assignment can be done independently, i.e., we do not need to design the processes to suit a particular hardware wiring of the processors. It is perfectly possible to run a whole set of processes on a single processor, and this is frequently done in order to test and debug programs.

9.2.2 Two Elementary Processes

There are two elementary processes in the language:

- **SKIP**

- **STOP**

Both processes are of vital importance to the philosophy of the language. In the previous section we dealt with the process concept. A process can start, perform some action and then terminate. The language also contains the **SKIP** process to provide for an empty action, i.e., a process which starts, but performs no action and immediately terminates (for the processes discussed in the previous section the action was non-empty). Of course the **SKIP** process can be used as a fill-in for later replacement by a process which actually does some action. However, its primary function is to be an empty action, and therefore it serves as more than a mere fill-in. We shall see later that such a process is indeed necessary.

The **STOP** process is the complement of the **SKIP** process. The **STOP** process starts but never terminates. This can be useful for handling errors while a program is being developed.

With the two elementary processes we have covered all the possibilities for the action. We have the empty action, the action with a finite number of steps, and the infinite (unending) action.

9.2.3 A Trivial Example

Let us make a short break in the exposition and look at a complete but trivial example of an executable Occam program. The example is the infamous "Hello World", which in Occam reads as follows:

```
PROC first(CHAN OF ANY keyboard, screen)

  -- First example program  --

  #USE "userio.tsr"
  #USE "megamath.tsr"

  SEQ
    newline(screen)
    write.text.line (screen, "hello world!")

    newline(screen)
    write.full.string(screen, "type any to return")
    INT any:
    read.char(keyboard, any)
:
```

The first statement in the Occam program is the procedure declaration. As in Fortran or other languages we need to declare some means for the communication with the outside world. This is done above with the *screen* and *keyboard* definitions. Following the procedure heading is the declaration part, which in our case consists of only one variable which will be declared later on. We also need to declare the libraries which the system uses to find such routines as *newline*. After we have printed out the text "hello world", we keep the output on the screen from disappearing by requiring the program to wait until some character is entered via the keyboard (otherwise the program would end immediately and the message would disappear from the screen). The ":" ends the Occam program.

9.2.4 Repetition and the Conditional

In the previous sections we have introduced some basic building blocks of the language. One building block is the process, which starts with a declaration part where all constants and variables and the names and bodies of other processes (procedures and functions) are declared. This is followed by the process body where some action is performed. The action can be a single statement such as an assignment, an empty process (**SKIP**) or the **STOP** process. The action, if it is a compound action, must either be performed in sequential (**SEQ**) order or in parallel (**PAR**). In this section we introduce two more basic building blocks which control the flow of execution of the actions.

a) The Repetition

To a large extent computational science lives from repetition. In a Monte Carlo algorithm [9.15–18], for example, we generate many configurations in order to obtain averages of quantities. The same applies to molecular dynamics simulations [9.17,19], where many integration steps must be performed. Consequently it is clear that we need a language element or construct allowing the repeated execution of a block of code, as in the following example:

loop $mcs = 1$ **to** $mcsmax$
    generate.configuration

The statements within the block labelled *generate.configuration* are executed exactly *mcsmax* times.

    Such a repetition can have two different meanings, although there is no sharp boundary between them. The first meaning is that we want to *iterate* something, where the iteration depends on the index as a counter

    a := 0
    loop $i = 1$ **to** n
      $a := a + f(i)$

or as a parameter

    loop $i = 1$ **to** n
      $a[i] := b[i] + c[i]$

    In Occam there are several constructs allowing the repetition in this sense. One also has the possibility of building a loop using a **WHILE** construct

    $mcs := 0$
    WHILE $mcs < mcsmax$
      SEQ
        generate.configuration
        $mcs := mcs + 1$

More generally, the **WHILE** allows for termination conditions other than the fixed iteration of the block. An example is given below of the Newton iteration for the determination of a zero of a function.

Example

Recall that the Newton method to determine a zero of a function $f(x)$ is given by

$$x_{i+1} = \Phi(x_i) \quad ,$$

$$\Phi(x) = x - \frac{f(x)}{f'(x)} \quad .$$

The termination condition of the Newton iteration method for finding a zero of a function $f(x)$ can be formulated as

$$\frac{x_i - x_{i+1}}{x_i} \le \epsilon \quad , \qquad i < maxit \quad .$$

A loop for the determination of a zero of $f(x)$ could be written as

```
i := 0
WHILE i < maxit AND abs(d) < eps * x
  Block
    generate.new.x
    i := i + 1
```

Since the termination depends upon a Boolean expression the possibility exists that the program remains in an endless loop:

```
WHILE TRUE
  action
```

Physically the termination condition can come from "outside", although from the logical point of view the condition is set from within the program. By "outside" we mean that a message over a channel is received which effects the Boolean expression.

Example

Often a process should run on a transputer until a finish signal is received from some designated master process. The corresponding process on the slave processor would look like

```
WHILE (terminate = FALSE)
  PAR
    process.to.accept.messages.and.commands
    process.to.work
```

If one of the messages accepted sets *terminate* to **TRUE**, then the process will stop.

We have already discussed the **SEQ** construct, within whose scope statements are executed in the order in which they appear. A replicated **SEQ** is like a loop where all the statements within its scope are executed several times:

SEQ $mcs = 1$ **FOR** $mcsmax$
  generate.configuration

It serves as an abbreviation for writing the process *generate.configuration* $mcsmax$ times. However, some care is required, because the index mcs serves as a label for the processes and therefore the index mcs and the **FOR** do not have the same meaning as they would have in Fortran or Pascal. This idea of the index being a parametrization of a process is the second meaning which the repetition can have. As such, the number of repetitions of the block must be fixed and the loop can be characterized as unconditional.

The idea may become more apparent when we note that we can also replicate the **PAR**. The statements

PAR $mcs = 0$ **FOR** $mcsmax$
  generate.configuration

mean that we want to repeat the process *generate.configuration* a number $mcsmax$ times but that we do not care what order they are carried out in. All the processes can be run in parallel and the index serves as a label or parametrization to identify the processes which could be assigned to $mcsmax$ different processors (how this is implemented in practice is discussed in Sect. 9.2.9). Again, it should be emphasized that the number of processes (i.e., $mcsmax$) must be known before the execution of the loop can proceed, and its value cannot be changed. Consequently the **PAR** is quite useful for the parallelization of replication algorithms since a single process can be copied many times and run concurrently on several processing elements with different parameters.

Example

For many algorithms we need to communicate results or messages from one process to another. The processes can be on different processors, on the same processor, or both these alternatives. If there is a message to be passed between processes then an action is performed corresponding to the message. Supposing that the messages come from the four links of the transputer, we could formulate an algorithm as

PAR $i = 0$ **FOR** 4
  SEQ
    link[i] ? message
    process.message

b) The Conditional

In Occam one must enumerate all the possible alternatives of a conditional – it does not suffice to give just one condition. For example, the condition

$$random < transition.prob(delta)$$

may happen to be false. In other languages if there is no other condition supplied then the conditional terminates. It is implicitly assumed that execution is continued with the next statement *after* the conditional. However, in Occam what should happen must be made explicit by the provision of alternative actions for every possible outcome of a conditional construct.

The **IF** construct is one realization of the concept of choice. It is not just a simple branching point in the execution of a program where the flow of the execution of statements is controlled by an **IF**, rather the **IF** in Occam offers us the choice of a process from a list of possibilities. One of the many possibilities is selected.

A conditional consists of two lists. The first list gives the conditions $L_c = \{condition_0, ..., condition_{n-1}\}$ and the second list gives the processes $L_p = \{process_0, ..., process_{n-1}\}$. Associated with each condition is a process

$$L_c \rightarrow L_p \quad .$$

Explicitly the **IF** construct is written as

IF
    $condition_0$
      $process_0$
    $\vdots$
    $condition_{n-1}$
      $process_{n-1}$

If a condition is true then the associated process is executed. After the associated process terminates the entire conditional terminates. The order of the conditions is thus crucially important. The list is worked through sequentially starting with the condition at the top and the process associated with the *first* condition encountered which gives the result **TRUE** is executed. This provides the possibility of having conditions which are not mutually exclusive.

Perhaps this point becomes a little clearer when we consider what possibilities exist for a conditional. A conditional is usually preceded by the evaluation of some expression for which at least three possible senarios can be listed:

- Comparison of the result to a given value

- Classification of the result

- Relation to other results

The first in this list is the usual **IF** ... **THEN** ... **ELSE** situation, and in the second a **CASE** construct applies. It is the third possibility where perhaps the largest number of logical alternatives present themselves.

The following construct is not only valid but also necessary in many cases:

TRUE
  SKIP

This makes it explicit that in the case that none of the other conditions apply the empty process is to be executed in order to terminate the conditional. If none of the conditions can be satisfied then the program does not proceed any further.

Example

Consider the spin exchange dynamics (Kawasaki dynamics) [9.20] for the Monte Carlo simulation of an Ising model with a conserved number of up and down spins. A lattice site is first chosen at random and then one of its q neighbours is chosen. If the two selected spins are of opposite sign, they are considered for an exchange. The selection of one of the q neighbours is done with the **IF** construct:

$neighbour \leftarrow random\{1, ..., q\}$
IF
  $neighbour = 1$
    SEQ
      $x.new := x + 1$
      $y.new := y$
  $neighbour = 2$
    SEQ
      $x.new := x$
      $y.new := y + 1$
  $neighbour = 3$
    SEQ
      $x.new := x - 1$
      $y.new := y$
  $neighbour = 4$
    SEQ
      $x.new := x$
      $y.new := y - 1$

Logically the **IF** construct appears as a **SEQ** with *guarded commands* [9.21]. The guard is a Boolean expression preceding a block and the block can be executed only if the expression is true. The block with the guard acts as a **SKIP** if the Boolean expression is false. The guards will be discussed again shortly in connection with the **ALT** construct.

The difference between the **IF** and the **SEQ** is the termination condition. Whereas the **SEQ** with guarded commands terminates only when *all* processes within its scope have terminated, the **IF** construct terminates after one guard is satisfied and the associated process has terminated.

The list of conditions can be quite long, in which case it makes sense to abbreviate it. Suppose we want to search a vector for a specific element *search*. This can be done by replicating the **IF** construct as follows:

$found := -1$
IF $i = 0$ **FOR** n
  $vector[i] = search$
    $found := i$

If the element *search* is not included in the vector the program will halt. To prevent this we can embed the replicated **IF** in another **IF**

$found := -1$
IF
  IF $i = 0$ **FOR** n
    $vector[i] = search$
      $found := i$
  TRUE
    SKIP

In the case that the search element is found in the vector, the **IF** branches out of the **FOR** loop.

Problems

9.1 Write an algorithm which switches the flow of data by reading a command from a link. Depending on the command, one of the three other links is selected and all other data coming in over the first link are directed to the chosen link until a termination marker is read.

9.2 Design an algorithm which adds two vectors $a[i]$ and $b[i]$ component-wise into a third vector $c[i]$. Here only two processors should be used.

9.3 What is the outcome of the following?

$n := 0$
SEQ $i = 0$ **FOR** n
  ...

9.4 Can the following be a valid part of an Occam program?

IF
  $test = 0$
    PAR
      ...
  $test()0$
    SEQ
      ...

9.2.5 An Occam Program for the Ising Model

Let us interrupt the introduction of new language elements and try to design a program with the language elements now available. At this stage we will not try to be too ambitious and will restrict ourselves to a single processor. Thus there is no attempt made here to use the concurrency. The program for the Ising model which is given below is basically a direct translation of a Fortran program into Occam. It serves as a starting point for more advanced programs using the new features of the language that will be introduced in subsequent sections, and we will give a truly parallel version of a Monte Carlo simulation of the Ising model in Appendix A.

Example

In this example we give a straightforward Monte Carlo simulation of the two-dimensional Ising model. Since this will be the first complete non-trivial Occam program in this book we will describe the parts of the program in some detail. As a first step all variables are declared (notice that in Occam counting starts with zero instead of one):

```
--- declaration part of the program ---

PROC two.d.ising(CHAN OF ANY keyboard, screen)

-------------------------------------------------------
--                                              --
--                2D ISING PROGRAM               --
--                                              --
--            on a single transputer             --
--                                              --
-------------------------------------------------------


  VAL two            IS 2.0 (REAL32) :
  VAL zero           IS 0.0 (REAL32) :
  VAL jkt            IS 0.440688 (REAL32) :

  [60][60] INT lattice     :
```

```
[60] INT im              :
[60] INT ip              :

[10] REAL32 t.prob        :


INT32 seed  :
INT L,mcsmax,discard :
INT linear, mcs.steps :
INT imy,ipy :
INT imx,ipx :
INT ici, ien :
INT iseed, cc, mag :
INT dummy :
INT StartTime,EndTime, TimeUsed, TimeCum :

REAL32 lattice.size :
REAL32 magnet, av.mag, av.mag.sqr, av.mag.four :
REAL32 cumulant, susceptibility :
REAL32 counter :
REAL32 exp, exp1            :
REAL32 t, ttc                    :
REAL32 ran                  :
REAL32 abs.mag, rmag, rcum               :
```

The procedure heading may look a little different for other installations. At the installation where this program was run we needed to define two channels *keyboard* and *screen* to connect the program with the outside world. Additional channels may need to be declared to perform structured input/output operations and it is left to the reader to fill-in the necessary details. The generic names *read* and *write* are used within the program to perform the input/output operations.

At the next step all the variables need to be initialized. Two arrays *im[L]* and *ip[L]* are declared to incorporate the periodic boundary conditions and are set such that the boundaries are identified with each other. The transition probabilities are set [the critical coupling for the two-dimensional Ising model is [9.22] $J/k_B T_c = (\log(1 + \sqrt{2}))/2$] and the lattice is initialized with all spins pointing in one direction.

```
---   initialization part of the program   ---

SEQ
```

```
newline(screen)
write(screen,"  2D Ising: ")
write(screen,"    program version 1.0")
newline(screen)

write(screen,"linear lattice size L        : ")
read(keyboard,screen,L,cc)
write(screen,"Monte Carlo Steps            : ")
read(keyboard,screen,mcsmax,cc)
write(screen,"discard Monte Carlo Steps    : ")
read(keyboard,screen,discard,cc)
write(screen,"Input seed                   : ")
read(keyboard,screen,iseed,cc)
write(screen,"temperature ttc              : ")
read(keyboard,screen,ttc,cc)
newline(screen)
newline(screen)


t             := jkt / ttc
seed          := INT32 iseed
mag           := - ((L * L) / 2)
av.mag        := 0.0 (REAL32)
av.mag.sqr    := 0.0 (REAL32)
av.mag.four   := 0.0 (REAL32)
lattice.size  := REAL32 ROUND (L*L)
counter       := 0.0 (REAL32)

SEQ i=0 FOR L
  SEQ  j=0 FOR L
    lattice[i][j] := -1


SEQ i=0 FOR L
  PAR
    im[i] := i - 1
    ip[i] := i + 1

im[0]   := L - 1
ip[L-1] := 0


SEQ i=0 FOR 10
  SEQ
```

```
exp := REAL32 ROUND  (i-5)
exp1 := - ((exp * t) * two)
EXPP(exp, exp1)
t.prob[i] := exp
```

The main part of the program contains the Monte Carlo sweeps through the lattice. There are two sequential loops over the two coordinates and an **IF** conditional to determine whether the Monte Carlo move is accepted or rejected. RANP is a generic procedure returning a uniformly distributed random number from the interval between zero and one.

```
--------------------------------------------------
--        M O N T E   C A R L O   P A R T    --
--------------------------------------------------

SEQ  mcs=1  FOR mcsmax
  SEQ
    SEQ j=0 FOR L
      SEQ
        imy := im[j]
        ipy := ip[j]
        SEQ i=0  FOR L
          SEQ
            imx := im[i]
            ipx := ip[i]
            ici := lattice[i][j]
            ien :=  lattice[imx][j] + lattice[ipx][j]
            ien := (lattice[i][imy] + lattice[i][ipy])
                    + ien
            ien := (ici * ien) + 5
            RANP(ran, seed)
            exp := t.prob[ien]
            IF
              ran < exp
                SEQ
                  lattice[i][j] := - ici
                  mag           := mag - ici

              ran >= exp
                SKIP
```

After the set number of sweeps *mcsmax* have been performed the averages over the various quantities are done and the results are sent to the screen or disk.

```
---    Analysis part of the program  ---
      IF
        mcs >= discard
          SEQ
            counter      := counter + 1.0 (REAL32)
            rmag         := REAL32 ROUND  (mag * 2)
            ABSP(magnet,rmag)
            av.mag       := av.mag + magnet
            magnet       := magnet * magnet
            av.mag.sqr   := av.mag.sqr + magnet
            magnet       := magnet * magnet
            av.mag.four  := av.mag.four + magnet
        mcs <  discard
          SKIP

  av.mag       := av.mag / counter
  av.mag.sqr   := av.mag.sqr / counter
  av.mag.four  := av.mag.four / counter

  susceptibility := av.mag.sqr - (av.mag * av.mag)
  susceptibility := susceptibility / lattice.size

  cumulant := (3.0 (REAL32) * (av.mag.sqr * av.mag.sqr))
              - av.mag.four
  cumulant := cumulant / (3.0 (REAL32) *
              (av.mag.sqr *  av.mag.sqr))

  av.mag := av.mag / lattice.size

  write(screen,"Monte Carlo Steps : ")
  write(screen,mcsmax,4)
  newline(screen)
  write(screen,"magnetization     : ")
  write(screen,av.mag,0,0)
  newline(screen)
  write(screen,"susceptibility  : ")
  write(screen,susceptibility,0,0)
  newline(screen)
  write(screen,"cumulant          : ")
  write(screen,cumulant,0,0)
```

```
newline(screen)
```

:

The reader is recommended to code this program and to do some runs to familiarize him/herself with the language. Later on we will be able to use parts of this program for a parallel version of the Ising model simulation.

Problems

9.5 Time the sequential version of the Monte Carlo simulation program of the Ising model.

9.6 There are many tricks to speed up the program. One is to put as much arithmetic as possible into a single line. Try this with the above program and compare the timing with your previous result. What is the limiting factor for the speed in the above program?

9.7 **Q2R Cellular Automata:** It is easy to modify the above program for the Metropolis simulation to the case where energy is explicitly conserved. One such way of implementing a conserved energy simulation is the Q2R cellular automaton [9.23]. We allow only those flips where there are an equal number of up and down spins in the neighbourhood, i.e., the sum of the nearest neighbours is zero. It is, of course, necessary to set up an initial configuration with the desired energy. Rewrite the above program so that it simulates a Q2R cellular automaton. Does it fulfil ergodicity? Can you find periodicities?

9.8 **Creutz Algorithm:** There is another way to perform a simulation at almost conserved energy which was proposed by *Creutz* [9.24,25]. The reader is asked to consult the literature on this simulation method [9.17,26] and to modify the above program so that a simulation at quasi-constant energy can be performed.

9.9 To speed up the execution of one sweep through the lattice it is possible to reduce the dimension of the array of the spin variables by mapping the two fields of the array into one long field. Instead of the usual periodic boundary conditions we can now employ helical boundary conditions, where the neighbouring spins are defined to be a fixed distance away in the one-dimensional array. What is the speed-up?

9.2.6 More on Choices and Selection

A tight synchronization of the individual processors is for many algorithms often neither necessary nor desirable. Many parallel computer architectures (for example the ICL Distributed Array Processor – DAP [9.27,28], and the Connection Machine) only allow the lock-step mode, where each processor carries out basically the same step of a computation at the same time. However the transputer allows greater flexibility since each processor can carry out independent calculations and there is no enforced global synchronization. Each processor is free to run on its own, until there is a request to synchronize with another processor for data communication.

The **ALT** construct introduces great flexibility by providing the possibility of responding to messages as they arrive. In an **ALT** construct several processes, which are called *alternatives*, are listed and each possibility has a *guard* associated with it. The guards are input processes and only the alternative associated with the first received input is executed. This gives the possibility of selective communication and solves the producer–consumer problem by introducing a lock on a processor which blocks all in-coming synchronization requests for a data transfer.

An algorithm that runs on several transputers and where the task performed on each single processor is identical may still be asynchronous because the execution times could be different on each processor. Consider, for example, the task which involves the determination of a pixel in a ray-tracing algorithm, in which one pixel is assigned to each processor. The number of iterations each processor has to do before the pixel element is calculated depends on the input parameters that the processor receives. Once the pixel is ready the result is sent via a link to another processor. Since the execution time can vary, the results will arrive at the links at different times. With the **ALT** we can write

```
    ⋮
ALT
  link1  ?  result
    supply.result.to.other.process
  link2  ?  result
    supply.result.to.other.process
```

In this construct the process to be executed will correspond to the link which receives a result first and the **ALT** terminates as soon as the process associated with the condition that *first* became true terminates. If two or more of the guards are ready at the same time then one of the alternatives is arbitrarily chosen to be executed. As in the case of an **IF** statement, it is also possible to have a process guarded by a **SKIP** and such a process is always ready.

In addition, the **ALT** construct allows the selective execution of processes depending on a Boolean expression and on the state of a channel, i.e. an event. Although this is much like an **IF**, one of the fundamental differences is that the

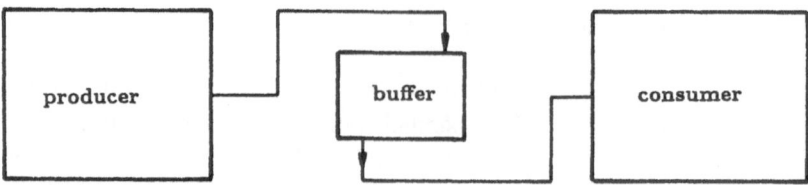

Fig. 9.7. Decoupling by a buffer of producer and consumer processes that operate at different speeds

conditions of the **IF** are worked through in sequential order. With this concept of a *guarded command* one can exclude some processes from being executed even though a message arrives over the channel. It is perhaps best to illustrate this with an example.

Example

Buffers (Fig. 9.7) provide a means of decoupling processes so that they can run asynchronously. Consider the situation where we have two processes, one that produces data and another that consumes the data. If the two processes work at different speeds then a lock-step mode would force them both to work at the speed of the slowest. However, a buffer enables us to decouple them in the following way. Suppose the producer operates with a faster speed than the consumer. As long as there is space in the buffer the producer deposits the results there and the consumer can take the results when they are required. Thus as soon as the producer has finished its tasks for one consumer it is free to go on to work for another without waiting for the first consumer to collect the results, since this can now be done at any time from the buffer.

    The kind of buffer which we want to look at here is a *ring buffer* of the type illustrated in Fig. 9.8, whose process could be designed as follows (we also assume that there are channels *from.producer*, *from.consumer* and *to.consumer* defined, the protocol for which will be discussed shortly):

head

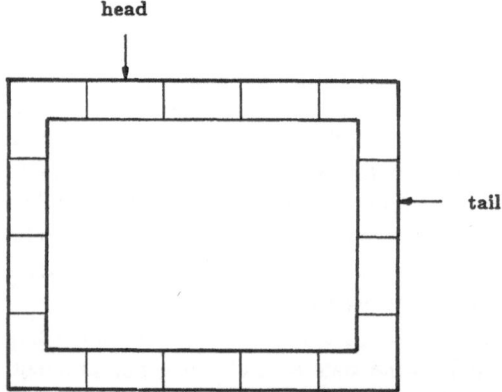

tail

Fig. 9.8. Ring buffer for the decoupling of a producer and a consumer

```
[buffersize] INT buffer :
INT head, tail, size, request :
head := 0
tail := 0
size := 0
WHILE still.go
  ALT
    size < buffersize & from.producer   ?   buffer[tail]
      SEQ
        size := size + 1
        tail := (tail + 1) REM buffersize
    size > 0 & from.consumer   ?   request
      SEQ
        to.consumer   !   buffer[head]
        size := size − 1
        head := (head + 1) REM buffersize
```

REM is a language element which allows us to carry out modulo operations. In this example, the variables *head* and *tail* can only take on values 0 to *buffersize-1*.

The guard locks out the producer in the case that the buffer is full and no data can be put into the buffer. As soon as the consumer has taken out data from the buffer the guard *size < buffersize* is true and a message can be received. On the other hand, if there are no data items in the buffer the consumer is locked out via the guard *size > 0*.

It is important to note that a **PAR** cannot be used here to replace the **ALT**. The **ALT** terminates as soon as the process associated with the condition that came true terminates. The **PAR** terminates only after every process within its scope has terminated.

We can also replicate the **ALT**, and the first example in this section could be written in abbreviated form as

```
    ⋮
ALT i = 0 FOR 2
  link[i]   ?   result
    supply.result.to.other.process
```

Problems

9.10 Write an algorithm which routes messages from the four input links to a single output link.

9.11 **The Dining Philosophers:** There are five philosophers who sit in a room and meditate. When one of them is hungry, he enters a dining room. On the circular table in the dining room there is a bowl of rice in the centre and a total of five chopsticks, one at each of the five sitting places. The hungry philosopher seats himself and tries to take the chopstick on his right hand side. He then tries to take the one on his left hand side. Only when he has both chopsticks can he start to eat. After finishing eating he puts down the chopsticks and leaves the room. Write a parallel algorithm for the problem which prevents a dead-lock.

9.12 **Multiplexer:** Write a procedure which multiplexes many software channels over one hardware channel. The multiplexer should be capable of input and output operations.

9.13 **Crossbar Switch:** Write a procedure which solves the crossbar switch problem, i.e., m producers must be connected to n consumers such that each producer can deliver a message to any of the n consumers.

9.2.7 Further Language Elements

a) Giving Priority

The previous sections dealt with the central language elements, which needed to be discussed separately because of their conceptual importance. This does not mean that the elements discussed in this section are less worthy, but rather that they form a second conceptual level.

In a number of algorithms it happens that a single processor executes several processes in parallel, of which some are more important for the efficient execution of the algorithm than others. An example that is frequently encountered is when a processor must through-route messages from one processor to another. Since the coordination number of a transputer is limited, and possibly the number of processes is larger than the number of processors (the mapping problem), there will be the need to pass messages to processors not directly connected via links. It is often important that the message is passed-on as soon as it arrives in order that the receiving processor does not wait idle during the delay and that there is no consequent degradation of the overall performance. Thus, if there are two processes running in parallel on a single processor and one of them involves some communication, then we would like to give the communication process a higher priority than the other process which is only carrying out computation.

The **PRI** is a tuning device for both the overall and the local performance. Higher priority on a single processor means an effectively larger time-slice given by the scheduler. If there is no explicit setting of priorities, all processes on a single processor will be given the same priority, and thus the same time-slice. In the above example of a communication process and a compute process the priority can be given to the communication by writing

PRI PAR
    communicate
    compute

With the **PRI PAR** construct the processes are given priority in the order in which they appear: the first process gets the highest priority, the second process the second highest priority and so on. Exactly how the time is distributed depends on the processes themselves.

The same principle applies to assigning priorities to the conditions and associated processes of the **ALT**. We saw earlier that if two guards became ready simultaneously then an arbitrary choice was made between them. However, with

PRI ALT $i = 0$ **FOR** n
    link[i] ? result
      process

if two or more processes are ready simultaneously the process that appears first in the list gets the highest priority and will be executed.

b) Channels and Protocol

So far we have assumed that the sender and the recipient of a message via a link know exactly what to send and what to expect. The channels which we have been using have the protocol

CHAN OF ANY *name* :

Recall that the header of the Ising program included two channels *keyboard* and *screen*. Messages passed with this protocol are byte sliced and the recipient must piece the message together. To avoid confusion, especially if the messages which are passed involve many different data types, one can specify a particular protocol for the channel

CHAN OF (REAL32, [10] INT) *channel.name* :

Here the messages which are passed via the *channel.name* consist of a **REAL32** number followed by an array of ten data items of type **INT**. The definition of the language allows for more elaborate protocols. There is, for example, a provision for a case-dependent protocol.

There is one channel in Occam which serves a single purpose and has a specific protocol. A channel declared with the **TIMER** attribute

TIMER *name* :

can only be read. Over this channel, which is an internal channel, the local time is passed and it is measured in *ticks*. The number of these ticks per second depends on the particular host transputer. The result of a query is assigned to a **INT** variable. A simple example may illustrate the use of the **TIMER**.

Example

In this example we want to time a routine, or a collection of statements, for their performance. For this purpose we set up the channel to read the clock of the processor, which we call *clock*. Over this channel we can read data of the type **INT** which gives the number of ticks passed. We can calculate the time used by reading *clock* at the beginning of the collection of statements with *TimeStart* and at the end with *TimeEnd* and then simply converting the number of ticks of *TimeUsed* into seconds or microseconds. The conversion factor depends on the particular processor and must be checked in the manual.

```
TIMER clock :
INT TimeStart, TimeEnd, TimeUsed :
SEQ
    clock   ?   TimeStart
    – do something in here –
    clock   ?   TimeEnd
    TimeUsed := TimeEnd − TimeStart
```

The timer channel can also be used to inhibit the processor for a given length of time. To this end, there is the language element **AFTER**, which produces a delay during which the processor is simply idle:

```
TIMER clock :
INT TimeIs, interval, delay  :
SEQ
    interval := 100
    clock   ?   TimeIs
    delay := TimeIs + interval
    clock   ? AFTER    delay
```

c) Abbreviations

Abbreviations often come in handy, for example we may reference a constant or a whole expression just by a name. A constant is defined, for example, by

VAL *pi* **IS 3.1415 (REAL32):**

Throughout the entire program the value of π can be used by referencing *pi*. This is just a simple and trivial example of the power of abbreviations. This could of course also be done within the program, but the advantage of the above abbreviation is that it can be placed before the body of the program and it then acts as a parameter statement.

Example

In this example we use the abbreviation facility to define a constant for the array size used in a program. This enables us to keep the code flexible for later readjustments.

VAL INT *n* **IS** 60 :
$$\vdots$$
[*n*][*n*] **INT** *old.array*

The possibility of abbreviating the use of arrays is quite useful. An array has multiple indices and in many computational science applications one of the indices is used in a loop while the others are not changed. In this case there is no need to use all indices since one can instead write

[*n*][*n*] **INT** *old.array*
$$\vdots$$
SEQ
$$\vdots$$
k := 6
new.array **IS** [*old.array*[*k*] **FROM** 0 **FOR** *n*]

The last statement is a declaration of a new array, which in this case is a vector with *n* components. Inside a loop this can result in a saving of time because the reference to elements of the *old.array* which are needed inside the loop is done via the *new.array*, and thus only one index computation needs to be done.

Example

Abbreviation is useful for the block transfer of data via channels. The full channel speed and throughput is reached only with block transfers. In this example we want to send a one-dimensional array across a channel. Instead of sending each element of the array we initiate a transfer of a segment of the array using the following abbreviation:

VAL INT *n* **IS** 60 :
$$\vdots$$
[*n*][*n*] **INT** *array*

$$\vdots$$

SEQ

$$\vdots$$

new.array **IS** [*array*[*k*] **FROM** 0 **FOR** *n*]
channel ! [*new.array* **FROM** *ptr* **FOR** *L.by.p*]

However, the power of the abbreviation is not limited to this and it is possible for complex expressions which are used many times in a program to be abbreviated.

Problems

9.14 What is wrong with the following excerpt from a program?

VAL *seed* **IS** 4711

$$\vdots$$

INT *seed* :

9.15 Write explicitly the code for the abbreviation of a two-dimensional array which is used inside two loops.

9.16 Use the above exercise to rewrite the Ising program employing all abbreviation facilities. Does the program run faster? Remind yourself what the limiting factor for the speed of the program was.

9.2.8 Arithmetic

Altogether there are five arithmetics in Occam! From a computational point of view this is sometimes advantageous. Often one knows exactly what values a particular variable can assume, such as in our example of the Ising model where the local energy can only assume a small number of integer values. For the two-dimensional Ising model the change in energy when a spin is updated can only take the values $-4, -2, 0, 2$, or 4. Having such small values means that there is no need to check for overflow after each arithmetic operation. If the only arithmetic to be done is to calculate the energy, which has a clear bound that is less than the largest possible integer, then all arithmetic can be done with an unchecked integer arithmetic, with resulting savings in execution time. The usual operations $+, -, *$, and $/$, for both integers and real, are all checked for an overflow. The addition

$a := 2147483647$
$b := 2$
$c := a + b$

would cause an overflow since the largest integer value for the data type **INT32** is $2^{31} - 1$. An overflow condition is not set if we use **PLUS** instead of +. Using **PLUS** the arithmetic is done modulo $2^{31} - 1$. The reader should try the following to see the result:

$a := 2147483647$
$b := 2$
$c := a$ **PLUS** b

Similarly there is a **TIMES** operator for modulo $2^{31} - 1$ multiplication and there is a **MINUS** operator. With these operators we have three arithmetics at our disposal:

- Checked integer

- Checked real

- Unchecked integer

Unfortunately the language has no priority concept for operators. In other programming languages there is usually a convention as to which operations within a single statement are carried out first. The statement

$a := a + b * c$

would be completely ambiguous in Occam, and in order to ensure that the multiplication is given a higher priority than the addition it is necessary to write

$a := a + (b * c)$.

In Occam one has to use brackets for every dyadic operator – and sometimes also for monadic operators. There is no implicit priority convention and to make an expression unambiguous all brackets need to be given even, as in the following example, where there is only one type of operator:

$a := a + ((b + c) + d)$.

Some operations that are normally associative are no longer so on computers, so that even summing up some values in a different order can give different results! With brackets the responsibility for how the operations are carried out is returned to you.

It almost goes without saying that mixing of data types in an arithmetic expression is not allowed in Occam. All data types of the expression must be identical to the data type of the receiving variable. This means that there is no implicit data conversion, i.e., rounding or truncation. The data type must be explicitly converted to the appropriate data type by either of the following:

- **ROUND**
- **TRUNC**

This is a bit bothersome since a mixing of data types often occurs when one uses random numbers to generate indices:

REAL32 r :
INT32 *index, L* :
SEQ
   $L := 15$
   $r \leftarrow random(0, 1)$
   index := **INT32 TRUNC** ($r*$ **(REAL32 ROUND** L))

Of course, we could declare L to be real, but sometimes this means introducing an additional variable of which we have to keep track.

The advantage of using the unchecked operators is a significant saving in time. All arithmetic operations in the innermost loop of our Ising program can be done with the unchecked operators.

Example

The following part of the Ising program is taken from our previous examples and the integer arithmetic has been replaced by the unchecked fast integer arithmetic.

```
--------------------------------------------------
--        M O N T E   C A R L O   P A R T    --
--------------------------------------------------

SEQ  mcs=1  FOR mcsmax
  SEQ
    SEQ j=0 FOR L
      SEQ
        imy := im[j]
        ipy := ip[j]
        SEQ i=0  FOR L
          SEQ
            imx := im[i]
            ipx := ip[i]
            ici := lattice[i][j]
            ien :=  lattice[imx][j] PLUS lattice[ipx][j]
            ien := (lattice[i][imy] PLUS lattice[i][ipy])
                    PLUS ien
            ien := (ici TIMES ien) PLUS 5
            RANP(ran, seed)
            exp := t.prob[ien]
```

```
IF
   ran < exp
     SEQ
       lattice[i][j] := - ici
       mag           := mag MINUS ici

   TRUE
     SKIP
```

Occam allows us also to do modulo arithmetic. In the expression

$a := (b + c) \text{ REM } p$

the addition is done modulo p.

As well as the arithmetic with integers and real numbers there is also an arithmetic over the space of $\{0,1\}$, which is especially useful for multi-spin-coding algorithms [9.29,30]. We have the operators $\wedge, \vee$ for the *and* and *or* operations as well as the $\times$ operator for the *exclusive or*. The complement is denoted in Occam by the symbol $\sim$. Furthermore, there are two shift operators $<<$ and $>>$ which can be used on integers, both of which are non-cyclic.

Problems

9.17 How do you exponentiate?

9.18 Is $a := -b * c$ a correct expression within the Occam language?

9.19 If you are still awake and have glanced at the examples you may be wondering if the authors have their arithmetic right! Where is the promised fifth arithmetic? As well as the arithmetic on numbers we can also do arithmetic on Boolean variables, which can take the values **TRUE** and **FALSE**. The operators are **AND, OR** and **NOT**. How do you express the exclusive-or?

9.20 **Multi-Spin-Coding:** The Ising model can be parallelized on a local level by both calculating the energies of several spins at the same time and also changing their values simultaneously. This is done by the multi-spin-coding algorithm [9.29,30]. Several spins, depending on the number of bits in the computer word, are stored in a single word. The operations on this word are then the bitwise logical operators that have been introduced in this section. Write an Occam program for a multi-spin-coding algorithm for the two-dimensional Ising model without global parallelism (single processor).

9.2.9 Placements

Programs can be developed on a single processor even if there are sections in the program which can run concurrently. On a single processor they are subject to a scheduling algorithm which mimics, through multi-tasking, a multi-processor environment. True concurrency is achieved by assigning the various tasks to different processors.

Unfortunately (or fortunately, depending on your point of view concerning how much control you want to have over how the computer carries out the task), there is as yet no automatic processor/process assignment. The assignment must be made by the user who must explicitly assign the processes to the processors and assign the channels to the links of the processor. In general the process/processor mapping is a notoriously difficult problem! For some algorithms the choice is natural, while for others certain optimization criteria inhibit a straightforward solution.

The following program that runs on a single processor

PAR
    $process.0$
    $\vdots$
    $process.(n-1)$

is easily modified to run concurrently on n processors by *placing* the processes on particular network processors identified by a logical numbering:

PLACED PAR
    PROCESSOR 0
        $process.0$
    $\vdots$
    PROCESSOR $(n-1)$
        $process.(n-1)$

In this particular example, of course, it is assumed that there are as many processors as processes. It is also possible to assign a number of processes to one network processor.

If the processes can be parametrized, as is often the case for example with homogeneous algorithms, then the **PLACED PAR** statement can be replicated:

PLACED PAR $i = 0$ **FOR** n
    PROCESSOR i
        $process(i)$

The configuration of the parallel program is achieved by the placement. Channels controlling the interaction between the processes on different processors now also need to be placed. On a single processor they were software channels, but in the multi-processor environment the channels must be assigned to physical channels: the links. Each transputer has four links with four input and four output

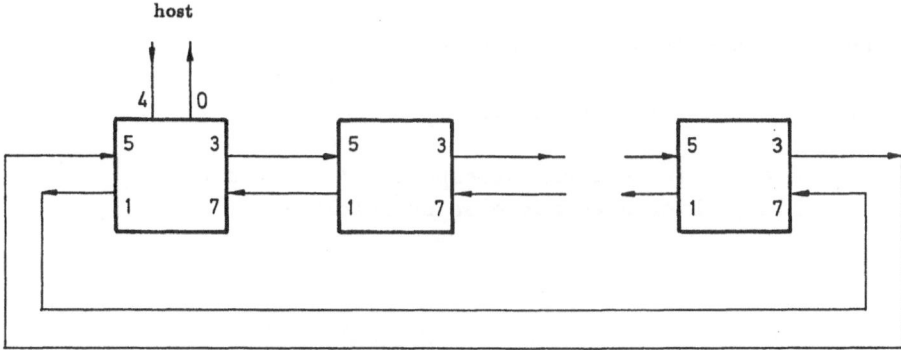

Fig. 9.9. A ring configuration for the placement of processors and communication links. The lines and arrows show the flow of information

channels, which are numbered as shown in Fig. 9.2. The placement of the links is done with the **PLACE ... AT** construct.

Example

In this example we want to configure a ring of n processes with $n - 2$ "inner" processes (*paris*) and two additional processes, one representing an "end" process (*paris.end*) and one connecting to the outside world (*paris.top*) as illustrated in Fig. 9.9. Data is read-in and results are read-out through *paris.top*, the generic processes *paris* pass on data and results to their neighbours, while the process *paris.end* is the end of the line and only sends results to its neighbour.

```
VAL  workers  IS  64:

PROC paris.top(CHAN OF ANY from.host, to.host,
                     reading.left,  writing.left,
                     reading.right, writing.right,
             INT processor )

  -- Process for network transputer connecting also
  -- to the outside world

:


PROC paris(CHAN OF ANY reading.left,  writing.left,
                    reading.right, writing.right,
             INT processor )

  -- Process for network transputer

:
```

```
PROC paris.end(CHAN OF ANY reading.left,  writing.left,
                            reading.right, writing.right,
               INT processor )

  -- Process for network transputer at the end of the
  -- line, which does not pass on information

:


[workers] CHAN OF ANY link.ring.1:
[workers] CHAN OF ANY link.ring.2:

PLACED PAR

  CHAN OF ANY to.host, from.host :

  PROCESSOR 0 T8
    PLACE from.host                    AT 6:
    PLACE to.host                      AT 2:
    PLACE link.ring.1[workers-1]       AT 4:
    PLACE link.ring.1[0]               AT 1:
    PLACE link.ring.2[0]               AT 5:
    PLACE link.ring.2[workers-1]       AT 0:

    paris.top(from.host,to.host,
              link.ring.1[workers-1],link.ring.2[workers-1],
              link.ring.2[0],        link.ring.1[0], 0)

  PLACED PAR i=1 FOR workers-2

    PROCESSOR i T8
      PLACE link.ring.1[i-1]           AT 4:
      PLACE link.ring.1[i]             AT 1:
      PLACE link.ring.2[i]             AT 5:
      PLACE link.ring.2[i-1]           AT 0:

      paris(link.ring.1[i-1], link.ring.2[i-1],
            link.ring.2[i],   link.ring.1[i], i)

  PROCESSOR workers-1 T8
    PLACE link.ring.1[workers-2]       AT 4:
    PLACE link.ring.1[workers-1]       AT 1:
```

```
PLACE link.ring.2[workers-1]        AT 5:
PLACE link.ring.2[workers-2]        AT 0:

paris.end(link.ring.1[workers-2], link.ring.2[workers-2],
          link.ring.2[workers-1], link.ring.1[workers-1],
          worker-1)
```

Problems

9.21 Write down the placement for a doubly connected ring with n elements, a binary tree with $n = 2^m$ elements and input/output between the father and son processsors, and a ternary tree.

9.22 A more demanding exercise is the placement of a two-dimensional grid, with and without periodic boundary conditions.

Appendices

A. A Parallel Ising Model Program

Most of the work necessary for an Occam program for a geometrically parallelized Ising program has been done already in the previous chapters. We only have to collect together the various pieces that we have developed so far. The strategy which we adopt for the geometric decomposition has been described in Chap. 7 on geometrically parallel algorithms.

For the sake of simplicity we work here in two dimensions and divide the lattice into strips. Each of the strips is handled by a single processor and in order to ensure detailed balance we divide the strips further into blocks. Each processor works on a block of its assigned strip such that blocks being updated on neighbouring processors do not have a common boundary. The easiest implementation of this scheme is a staggered assignment of blocks.

The overall ring communication harness and the placement of processes on processors was developed in Chap. 9. What we have to develop here is the generic process in the ring which performs the Monte Carlo steps and gathers the boundaries from the neighbouring processors.

Ising Model Program

The basis of this algorithm comes from the example in Sect. 9.2.5 on the Monte Carlo simulation of an Ising model. We only have to modify the array storing the spin orientation, and include the communication of the boundaries of the blocks. Note that we rotate the array storing the spin orientation in order to make use of the block transfer facility of Occam.

```
    --- declaration part of the program ---

PROC paris(CHAN OF ANY reading.left,  writing.left,
                        reading.right, writing.right,
           VAL INT processor)


-----------------------------------------------------
--                                                --
--                2D ISING PROGRAM                --
```

```
--                                              --
--      This is the generic process in the ring  --
--      for a lattice geometrically decomposed   --
--      into strips and blocks.                   --
--                                              --
---------------------------------------------------------


    VAL INT L          IS 256 :
    VAL INT p          IS 64  :

    VAL two               IS 2.0 (REAL32) :
    VAL zero              IS 0.0 (REAL32) :
    VAL jkt               IS 0.440688 (REAL32) :

    VAL INT L.by.p IS L/p :
    VAL INT Lp1    IS L.by.p + 1 :
    VAL INT Lp2    IS L.by.p + 2 :

    [Lp2][L] INT lattice :

    [L] INT im          :
    [L] INT ip          :
    [p] INT im.block    :
    [p] INT ip.block    :

    [10] REAL32 t.prob :

    INT32 seed  :

    INT LL :
    INT mcsmax, discard, iskip, iseed :
    INT linear, mcs.steps :
    INT imy,ipy :
    INT imx,ipx :
    INT ix,iy :
    INT ici, ien :
    INT cc, mag :
    INT offset,ptr :
    INT offset.p1,ptr1 :
    INT offset.p2,ptr2 :

    REAL32 lattice.size :
    REAL32 magnet, av.mag, av.mag.sqr, av.mag.four :
```

```
REAL32 flush.mag :
REAL32 cumulant, susceptibility :
REAL32 counter :
REAL32 exp, exp1                    :
REAL32 t,ttc   :
REAL32 ran                          :
REAL32 abs.mag, rmag, rcum                    :
```

Two pairs of channels are defined in the procedure heading. Each of the pairs handles the input/output of the boundaries from the neighbouring processor. The simulation results are also passed via these channels, although we could of course use one of the remaining two other links of the processor for the communication of the results (see problems). The name *processor* identifies the processor number on which the process is running. The starting block for the particular strip is derived from the value of *processor*, as is the seed for the random number generator.

The declaration of the variables and constants is almost the same as for the sequential algorithm. The only new features concern the declarations for the block division of the strip *im.block[p]* and *ip.block[p]*.

```
SEQ

    reading.left ?  LL;mcsmax;discard;iseed;ttc
    writing.right ! LL;mcsmax;discard;iseed;ttc

    iseed := iseed + processor

    seed         := INT32 iseed
    t            := jkt / ttc
    mag          := - ((L * L.by.p) / 2)
    av.mag       := 0.0 (REAL32)
    av.mag.sqr   := 0.0 (REAL32)
    av.mag.four  := 0.0 (REAL32)
    lattice.size := REAL32 ROUND (L*L)
    counter      := 0.0 (REAL32)

    SEQ i=0 FOR Lp2
      SEQ  j=0 FOR L
        lattice[i][j] := -1
```

```
SEQ i=0 FOR L
  SEQ
    im[i] := i - 1
    ip[i] := i + 1

im[0]    := L - 1
ip[L-1] := 0

SEQ i=0 FOR p
  SEQ
    im.block[i] := i - 1
    ip.block[i] := i + 1

im.block[0]    := p - 1
ip.block[p-1] := 0

SEQ i=0 FOR 10
  SEQ
    exp := REAL32 ROUND  (i-5)
    exp1 := - ((exp * t) * two)
    EXPP(exp, exp1)
    t.prob[i] := exp
```

To start the simulation the initial values of the simulation parameters are passed from one processor of the ring to another. The initialization follows as for the sequential algorithm, with periodic boundary conditions inside the strip being handled by the arrays *im[L]* and *ip[L]*. Periodic boundary conditions in the other direction are ensured by the ring structure of the communication harness. The only addition to the initialization part of the program compared to the sequential algorithm are the two arrays *im.block[p]* and *ip.block[p]* which handle the block structure within one strip.

```
--------------------------------------------------
--      M O N T E   C A R L O   P A R T      --
--------------------------------------------------


SEQ  mcs = 0  FOR mcsmax
  SEQ
    SEQ  block = 0  FOR  p
      SEQ

        offset    := (block + processor) REM p
        ptr       := offset * L.by.p
```

```
offset.p1 := ip.block[offset]
ptr1       := offset.p1 * L.by.p
offset.p2 := im.block[offset]
ptr2       := offset.p2 * L.by.p

-- read and write the boundaries
PAR
  PAR
    reading.right ? [lattice[Lp1] FROM ptr FOR L.by.p]
    reading.left  ? [lattice[0]   FROM ptr FOR L.by.p]
  PAR
    writing.right ? [lattice[L.by.p] FROM ptr1 FOR L.by.p]
    writing.left  ? [lattice[1]    FROM ptr2 FOR L.by.p]


SEQ j=ptr FOR L.by.p
  SEQ
    imy := im[j]
    ipy := ip[j]
    SEQ i=1  FOR L.by.p
      SEQ
        imx := im[i]
        ipx := ip[i]
        ici := lattice[i][j]
        ien :=  lattice[imx][j] + lattice[ipx][j]
        ien := (lattice[i][imy] + lattice[i][ipy])
                               + ien
        ien := (ici * ien) + 5
        RANP(ran, seed)
        exp := t.prob[ien]
        IF
          ran < exp
            SEQ
              lattice[i][j] := - ici
              mag           := mag - ici

          TRUE
            SKIP

IF
  mcs >= discard
    SEQ
      rmag          := REAL32 ROUND  (mag * 2)
      --  flush results
```

```
            reading.right ? magnet
            magnet := magnet + rmag
            writing.left ! magnet

    mcs < discard
      SKIP
```

:

The main part of the program is concerned with the Monte Carlo sweeps within a block. Before the spins in a block are updated the spins on the block's boundaries are communicated to its neighbours. After this communication has taken place the algorithm proceeds as before. In this particular example each new measurement of the magnetization of the strip is accumulated with the value received from the right neighbour, and is passed on to the neighbour on the left.

The interested reader is recommended to code this program, do some runs and then try some of the problems below, which expand upon the above algorithms.

Problems

A.1 Rewrite the above algorithm for the case of a three-dimensional lattice. Think about the best way to set up the lattice coordinates so that you can use the fast block transfer and/or the abbreviation facilities.

A.2 Can you invent another block scheme such that the communication of the boundaries can be done in parallel with the Monte Carlo part on a block? What is the saving in execution time over the above algorithm? Think of buffers!

A.3 **Q2R Cellular Automata:** In Sect. 2.1.2 and the problems in Sect. 9.2.5 concerning the sequential algorithm for the simulation of the Metropolis Ising model, the Q2R cellular automaton was introduced. For this algorithm one has to set up a configuration with the wanted energy as an initial configuration! Write a geometrically parallelized program for a Q2R cellular automaton.

A.4 **Multi-Spin-Coding:** The multi-spin-coding algorithm parallelizes, on a local level, the Metropolis Monte Carlo algorithm for the Ising model by calculating the energies of several spins at the same time and also changing them at the same time [A.1,2] (see Sect. 9.2.8 on arithmetic). Use this local parallelism and the global parallelism of a geometrically distributed lattice for a fast Metropolis Ising Monte Carlo program. How much faster is this algorithm compared to the other algorithms?

B. Random Number Generator

The quality of the random number generator is of utmost importance for the correctness of all numerical work that involves using random numbers, for example Monte Carlo simulations. It has been shown that correlations in a random number generator can lead to spurious results in simulations and completely unphysical predictions [B.1] Here we give an example of a random number generator based on the shift-bit-register algorithm [B.2–4] that has found widespread acceptance in applications of Monte Carlo simulations. The version of the algorithm given below is written in Occam [B.4].

Random Number Generator

Because Occam does not allow us to specify an array with variable dimensions, as for example in Fortran, it is necessary that the dimension of the array $x.f$ is fixed in the following program. When the algorithm is called it generates $n.f$ random numbers, to a maximum of *MaxNumbers*.

```
VAL  INT MaxNumbers  IS  20001 :
PROC r250 (VAL INT n.f, [Maxnumbers] REAL32 x.f, [251] INT32 m.f)

-- The array m.f has to be initialized with 250 INT32 random
-- numbers before the first call of r250 ( e.g., using the
-- built in RANP random number generator ). In this example
-- the maximum possible number of random numbers to be
-- generated in one call is taken to be 20000.
-- n.f is the actual number of random numbers
-- to be generated.

  INT iloop, num.of.loops, loop.rest, ind :
  INT irand, maxint :
  REAL32    rmax :

  SEQ
    maxint := #7EEEEEEE
    rmax := REAL32 ROUND maxint
    num.of.loops := n.f / 250
    loop.rest    := n.f REM 250
    irand        := 1
    SEQ iloop = 1 FOR num.of.loops
      SEQ
        SEQ ind = 1 FOR 147
          SEQ
            m.f[ind] := m.f[ind] >< m.f[ind+103]
            conversion (x.f[irand],m.f[ind])
```

```
      -- x.f[irand] := (REAL32 ROUND m.f[ind]) / rmax
      irand := irand + 1

  SEQ ind = 1 FOR 103
    SEQ
      m.f[ind+147] := m.f[ind+147] >< m.f[ind]
      conversion (x.f[irand],m.f[ind+147])
      -- x.f[irand] := (REAL32 ROUND m.f[ind]) / rmax
      irand := irand + 1

IF
  loop.rest = 0
    SKIP
  loop.rest <= 147
    SEQ ind = 1 FOR loop.rest
      SEQ
        m.f[ind] := m.f[ind] >< m.f[ind+103]
        conversion (x.f[irand],m.f[ind])
        -- x.f[irand] := (REAL32 ROUND m.f[ind]) / rmax
        irand := irand + 1
  loop.rest > 147
    SEQ
      SEQ ind = 1 FOR 147
        SEQ
          m.f[ind] := m.f[ind] >< m.f[ind+103]
          conversion (x.f[irand],m.f[ind])
          -- x.f[irand] := (REAL32 ROUND m.f[ind]) / rmax
          irand := irand + 1
      SEQ ind = 1 FOR (loop.rest - 147)
        SEQ
          m.f[ind+147] := m.f[ind+147] >< m.f[ind]
          conversion (x.f[irand],m.f[ind+147])
          -- x.f[irand] := (REAL32 ROUND m.f[ind]) / rmax
          irand := irand + 1
:
```

Problems

B.1 We saw in the problems in Sect. 9.2.5 concerning the Monte Carlo simulation
of the two-dimensional Ising model that the limiting factor for the speed
of the Monte Carlo algorithm is essentially the generation of the random

numbers. Determine the time necessary to draw one random number using the above algorithm.

B.2 One way to test an algorithm for the generation of random numbers is to check for the so-called d-space non-uniformity. Set up a d-dimensional array and use d successively generated random numbers to determine an entry in the array. If the random numbers are uncorrelated, then all the elements of the array will be visited. The number of unvisited elements should go to zero exponentially. Check this for the above algorithm.

C. A Parallel Molecular Dynamics Program

An efficient and straightforward way to parallelize a molecular dynamcis simulation of N interacting particles is to assign to each of the p available processors
N/p particles. If we have at our disposal exactly N processors we could do
the force calculations, which on a single processor requires of the order of N^2
operations, in N steps. This algorithm is discussed in Sect. 8.1 and is particularly appropriate for systems with long-range interactions between the particles.
Below we give a molecular dynamics program in Occam which is based on this
idea.

Molecular Dynamics Program

In this program N is the overall number of particles inside the computational
cell, which has a side length of $side$. The density of the particles inside the cell is
given by $density$. A velocity scaling is performed initially to set up the system
with a particular temperature $Tref$. The Lennard-Jones potential is cut off at
$rCutOff$, and the integration step is given by h.

The following is the main procedure, which collects the data generated by
the workers in the ring.

```
PROC main.md(CHAN OF ANY keyboard,screen)
  ------------------------------------------------------------
  --        m o l e c u l a r    d y n a m i c s
  --             microcanonical ensemble
  --     parallelized for p processors and n particles
  --       this is the <<host>> procedure for the ring
  ------------------------------------------------------------

  CHAN OF ANY  FromHead, ToHead :

  PLACE FromHead  AT  5:
  PLACE ToHead    AT  1:

  -- Definition of the constants for the simulation

  VAL INT N     IS 256  :
  VAL INT p     IS 2    :
  VAL INT nop   IS N/p  :
  VAL INT MaxPackets IS p - 1 :
  VAL INT iseed IS 4711 :

  VAL  density  IS  0.636     (REAL32) :
  VAL  side     IS  7.3834774 (REAL32) :
  VAL  Tref     IS  2.53      (REAL32) :
```

```
VAL  rCutOff  IS  2.5        (REAL32) :
VAL  h        IS  0.064      (REAL32) :

VAL  INT  MaxIntStep IS  50 :

-- other useful constants
VAL  zero    IS  0.0         (REAL32) :
VAL  half    IS  0.5         (REAL32) :
VAL  one     IS  1.0         (REAL32) :
VAL  four    IS  4.0         (REAL32) :
VAL  sixteen IS 16.0         (REAL32) :

INT dummy :
TIMER timer :

REAL32 real.time, time.int :
INT StartTime, EndTime, TimeUsed, TimeCum :

REAL32 hc,Trefc :
INT nopc :
INT32 seed :

REAL32 x,y,z :
[N] REAL32 vhx,vhy,vhz :

REAL32 sideh, hsq, hsq2 :
REAL32 Nreal,scale :
REAL32 N4, base,exp1 :
REAL32 disp :
REAL32 kEnergy,pEnergys,pEnergy, tEnergy,virial :
REAL32 rCutOff.square :
REAL32 Tscale :
REAL32 temperature, pressure :

INT iscale :

SEQ
  newline(screen)
  write.full.string(screen," Molecular Dynamics of Argon ")
  newline(screen)
  newline(screen)

  write.full.string(screen,"Initialization Start ")
  newline(screen)
```

```
sideh   := side * 0.5 (REAL32)
hsq     := h * h
hsq2    := hsq * 0.5 (REAL32)
rCutOff.square := rCutOff * rCutOff
Nreal   := REAL32 ROUND N

Tscale := sixteen / (Nreal - one)

Nreal   := REAL32 ROUND N
N4      := Nreal / four
ALOGP(base, N4)
exp1    := base * (1.0 (REAL32) / 3.0 (REAL32))
EXPP(scale, exp1)
disp    := side / scale
iscale := INT ROUND scale

TimeCum := 0

write.full.string(screen,"Set up lattice Start ")
newline(screen)
----------------------------------------------------
--  now the initial configuration is set up on an
--  fcc lattice inside of the box
----------------------------------------------------

SEQ  lg = 0  FOR 2
  SEQ  i = 0  FOR iscale
    SEQ  j = 0  FOR iscale
      SEQ  k = 0  FOR iscale
        SEQ
          x :=((REAL32 ROUND i)+((REAL32 ROUND lg)*half))*disp
          y :=((REAL32 ROUND j)+((REAL32 ROUND lg)*half))*disp
          z := (REAL32 ROUND k)* disp
          ToHead ! x;y;z

SEQ  lg = 1  FOR 2
  SEQ  i = 0  FOR iscale
    SEQ  j = 0  FOR iscale
      SEQ  k = 0  FOR iscale
        SEQ
          x :=((REAL32 ROUND i)+((REAL32 ROUND (2-lg))*half))*dis
          y :=((REAL32 ROUND j)+((REAL32 ROUND (lg-1))*half))*dis
          z :=((REAL32 ROUND k)+half) * disp
```

```
          ToHead ! x;y;z

-- assign normally distributed initial velocities
hc    := h
Trefc := Tref
nopc  := N
seed  := INT32 iseed
Maxwell(vhx,vhy,vhz,nopc,seed,hc,Trefc,Tscale)

SEQ  i = 0  FOR  N
  ToHead ! vhx[i] ; vhy[i]; vhz[i]


SEQ  IntStep = 0 FOR  MaxIntStep
  SEQ
    timer ? StartTime

    FromHead ? kEnergy;pEnergys;pEnergy

    timer ? EndTime
    TimeUsed     := EndTime - StartTime
    TimeCum      := TimeCum + TimeUsed

    kEnergy      := kEnergy * 24.0 (REAL32)
    pEnergy      := pEnergy * two
    tEnergy      := kEnergy + pEnergy
    temperature := Tscale * kEnergy

    write.full.string(screen,"step: ")
    write.int(screen,IntStep,4)
    write.full.string(screen,"Ekin = ")
    write.real32(screen,kEnergy,0,0)
    write.full.string(screen,"Epot = ")
    write.real32(screen,pEnergy,0,0)
    write.full.string(screen,"TEn = ")
    write.real32(screen,tEnergy,0,0)
    newline(screen)

real.time        := REAL32 ROUND TimeCum
real.time        := real.time / (REAL32 ROUND 15625)
time.int         := real.time / (REAL32 ROUND MaxIntStep)

write.full.string(screen,"Number of Integration Steps : ")
write.int(screen,MaxIntStep,4)
```

```
newline(screen)
write.full.string(screen,"time used was              : ")
write.real32(screen,real.time,0,0)
newline(screen)
write.full.string(screen,"time/integration step       : ")
write.real32(screen,time.int,0,0)
newline(screen)

write.full.string(screen,"ready")
keyboard ? dummy
```

:

The process for the basic worker in the ring is given below. Here the forces on the particles in the processor are computed first. The particle positions are then passed along the ring and each processor computes the forces which the passing particles exert on the particles belonging to the processor. After all particles have visited each processor, the positions and the velocities of all particles within each processor can be updated. The computed potential and kinetic energy are sent back to the main procedure for analysis.

```
PROC particle.head(CHAN OF ANY FromLeft,ToRight,FromHost,ToHost,
                   VAL INT processor)
------------------------------------------------------------
--          m o l e c u l a r   d y n a m i c s
--              microcanonical ensemble
--      parallelized for p processors and n particles
--          this is the <<head>> procedure in the ring
------------------------------------------------------------

-- Definition of the constants for the simulation

VAL INT N       IS 256  :
VAL INT p       IS 2  :
VAL INT nop     IS N/p :

VAL INT MaxPackets IS p - 1 :
VAL INT iseed   IS 4711 :

VAL  density  IS  0.636     (REAL32) :
VAL  side     IS  7.3834774 (REAL32) :
VAL  Tref     IS  2.53      (REAL32) :
VAL  rCutOff  IS  2.5       (REAL32) :
VAL  h        IS  0.064     (REAL32) :
```

```
VAL  INT  MaxIntStep IS  50 :

-- other useful constants
VAL  zero     IS  0.0        (REAL32) :
VAL  two      IS  2.0        (REAL32) :

REAL32 hc,Trefc :
INT nopc :
INT32 seed :

[nop] REAL32 x,y,z :
[nop] REAL32 vhx,vhy,vhz :
[nop] REAL32 force.x,force.y,force.z :
[nop] REAL32 pass.x, pass.y, pass.z :
[nop] REAL32 pass.vhx,pass.vhy,pass.vhz :
[nop] REAL32 got.x, got.y, got.z :

INT  IntStep :

REAL32 expo,rr, Nreal :
REAL32 sideh, hsq, hsq2 :
REAL32 kEnergy,pEnergys,pEnergy :
REAL32 kEnergyLeft,pEnergyLefts,pEnergyLeft :
REAL32 r, r148 :
REAL32 rCutOff.square :
REAL32 xd,yd,zd :
REAL32 kx,ky,kz :
REAL32 Tscale :

SEQ
  sideh  := side * 0.5 (REAL32)
  hsq    := h * h
  hsq2   := hsq * 0.5  (REAL32)
  rCutOff.square := rCutOff * rCutOff
  Nreal          := REAL32 ROUND N
  Tscale  :=16.0(REAL32)/(((1.0 (REAL32)*Nreal)-1.0(REAL32))

  -- get the initial positions of the particles from the
  -- host and distribute them further

  SEQ  i = 0  FOR  nop
    FromHost ? x[i] ; y[i]; z[i]

  SEQ  j = 0  FOR ((p - processor) - 1)
```

```
  SEQ  i = 0  FOR  nop
    SEQ
      FromHost ? pass.x[i] ; pass.y[i]; pass.z[i]
      ToRight  ! pass.x[i];pass.y[i];pass.z[i]

-- get the initial velocities of the particles from the
-- host and distribute them further

SEQ  i = 0  FOR  nop
  FromHost ? vhx[i] ; vhy[i]; vhz[i]

SEQ  j = 0  FOR ((p - processor) - 1)
  SEQ  i = 0  FOR  nop
    SEQ
      FromHost ? pass.vhx[i] ; pass.vhy[i] ; pass.vhz[i]
      ToRight  ! pass.vhx[i] ; pass.vhy[i] ; pass.vhz[i]

-- the forces on the particles are initially zero

SEQ i = 0  FOR  nop
  SEQ
    force.x[i] := zero
    force.y[i] := zero
    force.z[i] := zero

-----------------------------------------------------------
--  START OF THE ACTUAL MOLECULAR DYNAMICS PROGRAM
-----------------------------------------------------------

SEQ  IntStep = 0 FOR  MaxIntStep
  SEQ
    SEQ  i = 0  FOR  nop
      SEQ
        x[i] := x[i] + (vhx[i] + force.x[i])
        y[i] := y[i] + (vhy[i] + force.y[i])
        z[i] := z[i] + (vhz[i] + force.z[i])

    -- apply periodic boundary conditions --
    SEQ  i = 0  FOR  nop
      SEQ
        IF
          x[i] < zero
            x[i] := x[i] + side
          x[i] > side
```

```
            x[i] := x[i] - side
        TRUE
          SKIP
      IF
        y[i] < zero
          y[i] := y[i] + side
        y[i] > side
          y[i] := y[i] - side
        TRUE
          SKIP
      IF
        z[i] < zero
          z[i] := z[i] + side
        z[i] > side
          z[i] := z[i] - side
        TRUE
          SKIP


-- compute the partial velocities
SEQ  i = 0  FOR  nop
  SEQ
    vhx[i] := vhx[i] + force.x[i]
    vhy[i] := vhy[i] + force.y[i]
    vhz[i] := vhz[i] + force.z[i]


-- compute now the forces within

pEnergys := zero

SEQ  i = 0  FOR  nop
  SEQ
    force.x[i] := zero
    force.y[i] := zero
    force.z[i] := zero

SEQ  i = 0  FOR nop
  SEQ j = i + 1  FOR (nop - (i+1))
    SEQ

      xd := x[i] - x[j]
      yd := y[i] - y[j]
      zd := z[i] - z[j]

      -- apply the minimum image convention
```

```
IF
  xd < (-sideh)
    xd := xd + side
  xd >  sideh
    xd := xd - side
  TRUE
    SKIP
IF
  yd < (-sideh)
    yd := yd + side
  yd >  sideh
    yd := yd - side
  TRUE
    SKIP
IF
  zd < (-sideh)
    zd := zd + side
  zd >  sideh
    zd := zd - side
  TRUE
    SKIP

-- calculate the distance
r := ((xd*xd) + (yd*yd)) + (zd*zd)

IF
  r < rCutOff.square
    SEQ
      expo := - (6.0 (REAL32))
      POWERP(rr,r,expo)
      pEnergys    := pEnergys + rr
      expo := - (3.0 (REAL32))
      POWERP(rr,r,expo)
      pEnergys    := pEnergys - rr
      expo := - (7.0 (REAL32))
      POWERP(r148,r,expo)
      expo := - (4.0 (REAL32))
      POWERP(rr,r,expo)
      r148        := r148-( 0.5(REAL32) * rr)
      kx          := xd * r148
      force.x[i] := force.x[i] + kx
      force.x[j] := force.x[j] - kx
      ky          := yd * r148
      force.y[i] := force.y[i] + ky
```

```
                    force.y[j] := force.y[j] - ky
                    kz          := zd * r148
                    force.z[i] := force.z[i] + kz
                    force.z[j] := force.z[j] - kz
              r > rCutOff.square
                SKIP


------------------------------------------------
--  Now we compute the forces on the particles
--  which are handled by this processor. The
--  positions of the other particles are passed
--  along and the forces are accumulated
------------------------------------------------

SEQ  i = 0  FOR  nop
  SEQ
    pass.x[i] := x[i]
    pass.y[i] := y[i]
    pass.z[i] := z[i]

-- calculate forces on the particles within
-- the processor
pEnergy := zero

SEQ  packet = 0  FOR  MaxPackets
  SEQ
    -- send and receive the next packet
    SEQ  i = 0  FOR  nop
      PAR
        ToRight  ! pass.x[i];pass.y[i];pass.z[i]
        FromLeft ? got.x[i] ; got.y[i]; got.z[i]

    SEQ  i = 0  FOR nop
      SEQ j = 0  FOR nop
        SEQ
          xd := x[i] - got.x[j]
          yd := y[i] - got.y[j]
          zd := z[i] - got.z[j]

          -- apply the minimum image convention
          IF
            xd < (-sideh)
              xd := xd + side
            xd >   sideh
```

```
          xd := xd - side
      TRUE
        SKIP
  IF
    yd < (-sideh)
      yd := yd + side
    yd >  sideh
      yd := yd - side
    TRUE
      SKIP
  IF
    zd < (-sideh)
      zd := zd + side
    zd >  sideh
      zd := zd - side
    TRUE
      SKIP

  -- calculate the distance
  r := ((xd*xd) +( yd*yd)) + (zd*zd)

  IF
    r < rCutOff.square
      SEQ
        expo := - (6.0 (REAL32))
        POWERP(rr,r,expo)
        pEnergy    := pEnergy + rr
        expo := - (3.0 (REAL32))
        POWERP(rr,r,expo)
        pEnergy    := pEnergy - rr
        expo := - (7.0 (REAL32))
        POWERP(r148,r,expo)
        expo := - (4.0 (REAL32))
        POWERP(rr,r,expo)
        r148       := r148-(0.5(REAL32) * rr)
        kx         := xd * r148
        force.x[i] := force.x[i] + kx
        ky         := yd * r148
        force.y[i] := force.y[i] + ky
        kz         := zd * r148
        force.z[i] := force.z[i] + kz
    r > rCutOff.square
      SKIP
```

```
      SEQ  i = 0  FOR  nop
        SEQ
          pass.x[i] := got.x[i]
          pass.y[i] := got.y[i]
          pass.z[i] := got.z[i]


-------------------------------------------
-- this concludes the force calculation --
-------------------------------------------

-- compute the velocities
SEQ i = 0  FOR nop
  SEQ
    force.x[i] := force.x[i] * hsq2
    force.y[i] := force.y[i] * hsq2
    force.z[i] := force.z[i] * hsq2

SEQ i = 0  FOR nop
  SEQ
    vhx[i] := vhx[i] + force.x[i]
    vhy[i] := vhy[i] + force.y[i]
    vhz[i] := vhz[i] + force.z[i]

-- compute the partial kinetic energy
kEnergy := zero
SEQ  i = 0  FOR nop
  SEQ
    kEnergy := kEnergy+((vhx[i]*vhx[i])+
               ((vhy[i]*vhy[i])+(vhz[i]*vhz[i]))))

kEnergy := kEnergy / hsq

-- send back the results on the potential,
-- kinetic and virial energies

ToRight   ! kEnergy;pEnergys;pEnergy
FromLeft  ? kEnergyLeft;pEnergyLefts;pEnergyLeft
ToRight   ! kEnergy;pEnergys;pEnergy
FromLeft    ? kEnergyLeft;pEnergyLefts;pEnergyLeft

ToHost    ! kEnergyLeft;pEnergyLefts;pEnergyLeft

:
```

References

Chapter 2

2.1   K. Binder (ed.): *Monte Carlo Methods in Statistical Physics*, Topics Curr. Phys., Vol. 7, 2nd edn. (Springer, Berlin, Heidelberg 1986)

2.2   K. Binder, D.W. Heermann: *Monte Carlo Simulation in Statistical Physics: An Introduction*, Springer Ser. Solid-State Sci., Vol. 80 (Springer, Berlin, Heidelberg 1988)

2.3   D.W. Heermann: *Computer Simulation Methods in Theoretical Physics*, 2nd edn. (Springer, Berlin, Heidelberg 1990)

2.4   O.G. Mouritsen: *Computer Studies of Phase Transitions and Critical Phenomena*, Springer Ser. Comput. Phys. (Springer, Berlin, Heidelberg 1984)

2.5   M.H. Kalos: *Monte Carlo Methods* (Wiley, New York 1986)

2.6   N. Metropolis, A.W. Rosenbluth, M.N. Rosenbluth, A.H. Teller, E. Teller: J. Chem. Phys. **21**, 1087 (1953)

2.7   E. Ising: Z. Phys. **31**, 253 (1925)

2.8   D.J.E. Callaway, A. Rahman: Phys. Rev. Lett. **49**, 613 (1982)

2.9   G. Parisi, Wu Yongshi: Sci. Sin. **24**, 483 (1981)

2.10   S. Duane: Nucl. Phys. B **257**, [FS14], 652 (1985)

2.11   S. Duane, J. Kogut: Phys. Rev. Lett. **55**, 2774 (1985)

2.12   S. Duane, A.D. Kennedy, B.J. Pendleton, D. Roweth: Phys. Lett. B **195**, 216 (1987)

2.13   G.Y. Vichniac: Physica D **10**, 96 (1984)

2.14   Y. Pomeau, G.Y. Vichniac: J. Phys. A **21**, 3297 (1988)

2.15   H.J. Herrmann: J. Stat. Phys. **45**, 145 (1986)

2.16   M. Creutz: Phys. Rev. Lett. **50**, 1411 (1983)

2.17   S.L. Adler: Phys. Rev. D **23**, 2901 (1981)

2.18   M. Creutz: Phys. Rev. D **36**, 515 (1987)

2.19   G.G. Batrouni, G.R. Katz, A.S. Kronfeld, G.P. Lepage, B. Svetitsky, K.G. Wilson: Phys. Rev. D **32**, 2736 (1985)

2.20   J. Goodman, A.D. Sokal: Phys. Rev. Lett. **56**, 1015 (1986)

2.21   D. Kandel, E. Domany, D. Ron, A. Brandt, E. Loh: Phys. Rev. Lett. **60**, 1591 (1988)

2.22   A. Brandt: In *Third Copper Mountain Conference on Multigrid Methods*, ed. by S. McCormick (Dekker, New York 1988)

2.23   R.H. Swendsen, J.-S. Wang: Phys. Rev. Lett. **58**, 86 (1987)

2.24   U. Wolff: Phys. Rev. Lett. **62**, 361 (1989)

2.25   R.J. Glauber: J. Math. Phys. **4**, 294 (1963)

2.26   K. Binder, A.P. Young: Rev. Mod. Phys. **58**, 801 (1986)

2.27   J.D. Gunton, M. san Miguel, P.S. Sahni: In *Phase Transitions and Critical Phenomena*. Vol.8, ed. by C. Domb, J.L. Lebowitz (Academic, New York 1983)

2.28   A.N. Burkitt, D.W. Heermann: Europhys. Lett. **10**, 207 (1988)

2.29   K. Binder, D.W. Heermann: In *Scaling Phenomena in Disordered Systems*, ed. by R. Pynn, T. Skjeltrop (Plenum, New York 1985)

2.30 A. Milchev, K. Binder, D.W. Heermann: Z. Phys. B **63**, 521 (1986)
2.31 A. Baumgärtner, K. Binder, K. Kremer: Faraday Symp. Chem. Soc. **18**, 37 (1983)
2.32 A. Baumgärtner: J. Polym. Sci. C. Symp. **73**, 181 (1985)
2.33 H.J. Herrmann: Phys. Rep. **136**, 143 (1986)
2.34 A. Sadiq, K. Binder: Surf. Sci. **128**, 350 (1984)
2.35 N. Madras, A.D. Sokal: J. Stat. Phys. **50**, 109 (1988)
2.36 U. Wolff: Phys. Let. B **228**, 379 (1989)
2.37 K. Binder, D.W. Heermann, A. Milchev, A. Sadiq: In *Heidelberg Colloquium on Glassy Dynamics*, ed. by J.L. van Hemmen, I. Morgenstern, Lect. Notes Phys., Vol.275 (Springer, Berlin, Heidelberg 1987) p.154
2.38 R.G. Palmer: Adv. Phys. **31**, 669 (1982)
2.39 O.G. Mouritsen: Phys. Rev. B **32**, 1632 (1985)
2.40 B.J. Alder, T.E. Wainwright: J. Chem. Phys. **27**, 1208 (1957)
2.41 H.C. Andersen: J. Chem. Phys. **72**, 2384 (1980)
2.42 F.F. Abraham: Adv. Phys. **35**, 1 (1985)
2.43 D. Fincham, D.M. Heyes: Adv. Chem. Phys. **63**, 493 (1985)
2.44 D.J. Tildesley: In *Computational Physics*, ed. by R.D. Kenway, G.S. Pawley (SUSSP, Edinburgh 1987)
2.45 M.P. Allen, D.J. Tildesley: *Computer Simulation of Liquids* (Clarendon, Oxford 1987)
2.46 W.G. Hoover: *Molecular Dynamics*, Lect. Notes Phys., Vol.258 (Springer, Berlin, Heidelberg 1986)
2.47 G. Ciccotti, W.G. Hoover (eds.): *Molecular Dynamics Simulation of Statistical Mechanics Systems* (North-Holland, Amsterdam 1986) [Proc. of the Int'l School of Physics "Enrico Fermi" Course XCVII, Varenna, Italy 1985]
2.48 L.V. Woodcock: Chem. Phys. Lett. **10**, 257 (1970)
2.49 C.W. Gear: *Numerical Initial Value Problems in Ordinary Differential Equations* (Prentice-Hall, Englewood Cliffs, NJ 1966)
2.50 L. Verlet: Phys. Rev. **165**, 201 (1968)
2.51 R.W. Hockney: Methods Comput. Phys. **9**, 136 (1970)
2.52 W.C. Swope, H.C. Andersen, P.H. Berens, K.R. Wilson: J. Chem. Phys. **76**, 637 (1982)
2.53 G.D. Venneri, W.G. Hoover: J. Comput. Phys. **73**, 468 (1987)
2.54 P.P. Ewald: Ann. der Phys. **64**, 253 (1921)
2.55 S.G. Brush, H.L. Sahlin, E. Teller: J. Chem. Phys. **45**, 2102 (1966)
2.56 D.J. Adams: J. Chem. Phys. **78**, 2585 (1983)
2.57 D.W. Heermann, W. Klein, D. Stauffer: Phys. Rev. Lett. **49**, 1262 (1982)
2.58 M. Parrinello, A. Rahman: Phys. Rev. Lett. **45**, 1196 (1980)
2.59 J.A. Barker, D. Henderson: Rev. Mod. Phys. **48**, 587 (1976)
2.60 L. Verlet: Phys. Rev. **159**, 98 (1967)
2.61 R.W. Hockney, J.W. Eastwood: *Computer Simulation Using Particles* (McGraw-Hill, New York 1981)
2.62 J. Morales, F. Rull, S. Toxvaerd: Comput. Phys. Commun. **56**, 56 (1989)
2.63 D.C. Rapaport: Comput. Phys. Rep. **9**, 1 (1988)
2.64 G.C. Fox, S.W. Otto: Phys. Today **37**, 50 (May 1984)
2.65 T. Schneider, E. Stoll: Phys. Rev. B **13**, 1216 (1976); ibid. **17**, 1302 (1978)
2.66 G.S. Grest, K. Kremer: Phys. Rev. A **33**, 3628 (1986)
2.67 M. Creutz: Phys. Rev. Lett. **63**, 9 (1989)
2.68 D.W. Heermann, P. Nielaba, M. Rovere: Comput. Phys. Commun. **60**, 331 (1990)
2.69 F. Reif: *Fundamentals of Statistical and Thermal Physics*, (McGraw-Hill, New York 1965)

2.70 P. Nielaba, D.W. Heermann: Preprint (1990)
2.71 B. Dünweg, K. Kremer, G.S. Grest: Comput. Phys. Commun. 55, 269 (1989)
2.72 M.N. Barber: *Phase Transitions and Critical Phenomena*, Vol.8 (Academic, New York 1983)
2.73 M. Suzuki: Int. J. Magn. 1, 123 (1971)
2.74 K. Binder: In *Phase Transitions and Critical Phenomena*, ed. by C. Domb, M.S. Green (Academic, New York 1976)
2.75 C. Kalle: J. Phys. A 17, L801 (1985)
2.76 J.K. Williams: J. Phys. A 18, 49 (1985)
2.77 S. Wansleben, D.P. Landau: J. Appl. Phys. 61, 3968 (1987)
2.78 D.W. Heermann, R.C. Desai: Comput. Phys. Commun. 50, 297 (1988)
2.79 M.E. Fisher: In *Critical Phenomena*, ed. by M.S. Green (Academic, New York 1971) [Proc. of the 51st Enrico Fermi Summer School, Varenna, Italy 1970]
2.80 M.E. Fisher, M.N. Barber: Phys. Rev. Lett. 28, 1516 (1972)
2.81 P.W. Kasteleyn, C.M. Fortuin: J. Phys. Soc. Jpn. 26 (Suppl.), 11 (1969)
2.82 C.M. Fortuin, P.W. Kasteleyn: Physica (Utrecht) 57, 536 (1972)
2.83 W. Lenz: Phys. Z. 21, 613 (1920)
2.84 L. Onsager: Phys. Rev. 65, 117 (1944)
2.85 A.N. Burkitt, D.W. Heermann: Comput. Phys. Commun. 54, 201 (1989)
2.86 D. Stauffer: *Introduction to Percolation Theory* (Taylor and Francis, London 1985)
2.87 M.E. Fisher: Physics 3, 255 (1967)
2.88 A. Coniglio, W. Klein: J. Phys. A 13, 2775 (1980)
2.89 C.K. Hu: Phys. Rev. B 29, 5103 5109 (1984)
2.90 M. Sweeny: Phys. Rev. B 27, 4445 (1983)
2.91 D.W. Heermann, A.N. Burkitt: Physica A 162, 210 (1990)
2.92 J. Hoshen, R. Kopelman: Phys. Rev. B 14, 3428 (1976)
2.93 J. Kertesz: Private communication
2.94 M. Hasenbusch: Improved Estimators for a Cluster Update of O(n) Spin Models; Kaiserslautern Preprint TH – 23/89 (revised) (1989)
2.95 U. Wolff: Nucl. Phys. B 322, 759 (1989)
2.96 U. Wolff: Phys. Lett. B 222, 473 (1989)
2.97 Ch. Frick, K. Jansen, P. Seuferling: Phys. Rev. Lett. 63, 2613 (1989)
2.98 U. Wolff: Phys. Rev. Lett. 60, 1461 (1988)
2.99 U. Wolff: Nucl. Phys. B 300 [FS22], 501 (1988)
2.100 F. Niedermayer: Phys. Rev. Lett. 61, 2026 (1988)
2.101 R.G. Edwards, D. Sokal: Phys. Rev. D 38, 2009 (1988)

Chapter 3

3.1 C. Kalle, V. Winkelmann: J. Stat. Phys. 28, 639 (1982)
3.2 A.D. Kennedy, J. Kuti, S. Meyer, B. Pendelton: J. Comput. Phys. 64, 133 (1986)
3.3 D. Stauffer: *Introduction to Percolation Theory* (Taylor and Francis, London 1985)
3.4 T. Witten, L.M. Sanders: Phys. Rev. Lett. 47, 1400 (1981)
3.5 A. Baumgärtner, K. Binder, K. Kremer: Faraday Symp. Chem. Soc. 18, 37 (1983)
3.6 A. Baumgärtner: J. Polym. Sci. C. Symp. 73, 181 (1985)
3.7 K. Kremer, K. Binder: Comput. Phys. Rep. 7, 260 (1988)
3.8 B. Forrest, A. Baumgärtner, D.W. Heermann: Comput. Phys. Commun. 59, 455 (1990)
3.9 B.J. Alder, T.E. Wainwright: J. Chem. Phys. 27, 1208 (1957)

3.10 B.J. Alder, T.E. Wainwright: J. Chem. Phys. **31**, 456 (1959)
3.11 D.C. Rapaport: Comput. Phys. Rep. **9**, 1 (1988)
3.12 M. Eden: In *Proc. Fourth Berkeley Symp. on Math. Stat. and Prob.*, Vol.IV, ed. by F. Neyman (University of California Press, Berkeley, CA 1961)
3.13 P.L. Leath: Phys. Rev. B **14**, 5064 (1976)

Chapter 4

4.1   M.J. Quinn: *Designing Efficient Algorithms for Parallel Computers* (McGraw-Hill, Singapore 1987)
4.2   J.C. Wyllie: The Complexity of Parallel Computations. Ph.D Thesis, Dept. of Computer Science, Cornell University, Ithaca, NY (1979)
4.3   D. Knuth: *Fundamental Algorithms* (Addison Wesley, Reading, MA 1977)
4.4   C.L. Liu: *Elements of Discrete Mathematics*, 2nd edn. (McGraw-Hill, Singapore 1986)
4.5   A. Gibbons, W. Rytter: *Efficient Parallel Algorithms* (Cambridge Univ. Press, Cambridge 1988)
4.6   D. Nicol, F.R. Willard: J. Parallel Distrib. Comput. **5**, 404 (1988)
4.7   R. Friedberg, J.E. Cameron: J. Chem. Phys. **52**, 6049 (1970)
4.8   M. Creutz, L. Jacobs, C. Rebbi: Phys. Rev. Lett. **42**, 1390 (1979); Phys. Rev. D **20**, 1915 (1979)
4.9   R. Zorn, H.J. Herrmann, C. Rebbi: Comput. Phys. Commun. **23**, 337 (1981)
4.10 S. Wolfram: *Theory and Applications of Cellular Automata* (World Scientific, Singapore 1986)
4.11 M. Creutz: Phys. Rev. Lett. **50**, 1411 (1983)

Chapter 5

5.1   Hardware Reference Manual, CRAY Research (1984)
5.2   Hardware Reference Manual, Alliant
5.3   Hardware Reference Manual, Convex
5.4   S.F. Reddaway: DAP - a distributed array processor, in 1st Annual Symp. on Computer Architecture (IEEE/ACM), Florida, 1973 (IEEE, New York 1973)
5.5   W. Hillis: Sci. Am. 108 (June 1987)
5.6   Computing Surface: Hardware Reference Manual, Meiko, Bristol (1987)
5.7   Supercluster: Hardware Reference Manual, Parsytec, Aachen (1988)
5.8   Suprenum: Hardware Reference Manual, Suprenum, Bonn (1988)
5.9   R.W. Hockney, C.R. Jesshope: *Parallel Computers* (Hilger, Bristol 1981)
5.10 R.W. Hockney, C.R. Jesshope: *Parallel Computers 2* (Hilger, Bristol 1989)
5.11 K. Hwang, F.A. Briggs: *Computer Architecture and Parallel Processing* (McGraw-Hill, Singapore 1985)
5.12 BUTTERFLY: Hardware Reference Manual (1988)
5.13 C.L. Seitz: Commun. ACM **28**, 22 (1985)
5.14 J. Pattner: "Concurrent Processing: A New Direction in Scientific Computing" in Proc. National Computer Conference AFIPS, Vol.54 (AF IPS, Arlington, VA 1985)
5.15 V. Benes: *Mathematical Theory of Connecting Networks and Telephone Traffic* (Academic, New York 1965)
5.16 K. Binder, A.P. Young: Rev. Mod. Phys. **58**, 801 (1986)
5.17 J.H. Condon, A.T. Ogielski: Rev. Sci. Instrum. **56**, 1691 (1985)
5.18 K. Binder (ed.): *Monte Carlo Methods in Statistical Physics*, Topics Curr. Phys., Vol.7, 2nd edn. (Springer, Berlin, Heidelberg 1986)

5.19   K. Binder, D.W. Heermann: *Monte Carlo Simulation in Statistical Physics: An Introduction*, Springer Ser. Solid-State Sci., Vol.80 (Springer, Berlin, Heidelberg 1988)

5.20   D.W. Heermann: *Computer Simulation Methods in Theoretical Physics*, 2nd edn. (Springer, Berlin, Heidelberg 1990)

5.21   M.H. Kalos: *Monte Carlo Methods* (Wiley, New York 1986)

5.22   N. Metropolis, A.W. Rosenbluth, M.N. Rosenbluth, A.H. Teller, E. Teller: J. Chem. Phys. **21**, 1087 (1953)

5.23   A. Hoogland, J. Spaa, B. Selman, A. Compagner: J. Comput. Phys. **51**, 250 (1983)

5.24   R. Pearson, J.L. Richardson, D. Toussaint: J. Comput. Phys. **51**, 241 (1983)

5.25   G.S. Pawley, R.H. Swendsen, D.J. Wallace, K.G. Wilson: Phys. Rev. B **29**, 4030 (1984)

5.26   J. Beetem, M. Denneau, D. Weingarten: In IEEE Proc. of the 12th Annual Int'l Symp. on Computer Architecture (IEEE Computer Society, Washington, DC 1985)

5.27   J. Beetem, M. Denneau, D. Weingarten: J. Stat. Phys. **43**, 1171 (1986)

5.28   P. Bacilieri et al.: In *Computing in High Energy Physics*, ed. by L.O. Hertzberger, W. Hoogland (Elsevier, Amsterdam 1986)

5.29   N.H. Christ: J. Stat. Phys. **43**, 1061 (1986)

5.30   F. Hayot, H.J. Herrmann, J.-M. Normand: J. Comput. Phys. **641**, 380 (1986)

5.31   J. Zabolitzky: "KAPPES", Internal Report, Universität Köln (1985)

5.32   D.J. Tildesley: In *Computational Physics*, ed. by R.D. Kenway, G.S. Pawley (SUSSP, Edinburgh 1987)

5.33   W.G. Hoover: *Molecular Dynamics*, Lect. Notes Phys., Vol.258 (Springer, Berlin, Heidelberg 1986)

5.34   B. Dünweg, K. Kremer, G.S. Grest: Comput. Phys. Commun. **55**, 269 (1989)

5.35   Survey by the CCP5 Group. Information Quarterly, Daresbury (1989)

5.36   A.F. Bakker, C. Bruin, F. van Dieren, H.J. Hilhorst: Phys. Lett. A **93**, 67 (1982)

5.37   D.J. Auerbach, A.F. Bakker, T.C. Chen, A.A. Munshi, W.J. Paul: Mater. Res. Soc. Symp. Proc. **63**, 219 (1985)

5.38   D.J. Auerbach, W. Paul, A.F. Bakker, C. Lutz, W.E. Rudge, F.F. Abraham: J. Phys. Chem. **91**, 4881 (1987)

5.39   N. Neschen: "SUPERBUS"; Internal Report, Universität Köln (1989)

5.40   American National Standards Institute: *American National Standards Institute FORTRAN, X3.9-1978 (FORTRAN 77)* (ANSI, New York 1978)

5.41   SUPRENUM-FORTRAN: Suprenum-Report (Suprenum, Bonn 1988)

5.42   PAR-C Parsytec, Aachen (1988)

5.43   *Occam Programming Language* (Prentice-Hall, Englewood Cliffs, NJ 1984)

5.44   R. Steinmetz: *Occam 2* (Hüthig, Heidelberg 1987)

5.45   G. Jones: *Programming in Occam* (Prentice-Hall, Englewood Cliffs, NJ 1987)

5.46   K.C. Bowler, R.D. Kenway, G.S. Pawley, D. Roweth: *Occam 2 Programming Language* (Prentice-Hall, Englewood Cliffs, NJ 1984)

5.47   R.W. Hockney: Methods Comput. Phys. **9**, 136 (1970)

5.48   R. Friedberg, J.E. Cameron: J. Chem. Phys. **52**, 6049 (1970)

5.49   M. Creutz, L. Jacobs, C. Rebbi: Phys. Rev. Lett. **42**, 1390 (1979); Phys. Rev. D **20**, 1915 (1979)

5.50   R. Zorn, H.J. Herrmann, C. Rebbi: Comput. Phys. Commun. **23**, 337 (1981)

5.51   D. Barkai, K.M. Moriarty: Comput. Phys. Commun. **25**, 57 (1982)

5.52   W. Oed: Appl. Informatics **7**, 358 (1982)

5.53   S. Bakhai: IEEE Trans. C-30, 207 (1981)
5.54   J.W. Hong, K. Mehlhorn, A.L. Rosenberg: J. Assoc. Comput. Mach. **30**, 709 (1983)
5.55   S.S. Pink, Y. Wolfstahl: Int. J. Parallel Comput. **16**, 1 (1987)

Chapter 6

6.1   D.F. Rogers: *Procedural Elements for Computer Graphics* (McGraw-Hill, New York 1985)

Chapter 7

7.1   H.T. Kung: In *Advances in Computers*, Vol.19, ed. by M. Yovitts (Academic, New York 1980) pp.65-112
7.2   C.R. Askew, D.B. Carpenter, J.T. Chalker, A.J.G. Hey, D.A. Nicole, D.J. Pritchard: Comput. Phys. Commun. **42**, 21 (1986)
7.3   C.R. Askew, D.B. Carpenter, J.T. Chalker, A.J.G. Hey, M. Moore, D.A. Nicole, D.J. Pritchard: Parallel Comput. **6**, 247 (1988)
7.4   R.H. Swendsen, J.-S. Wang: Phys. Rev. Lett. **58**, 86 (1987)
7.5   P.W. Kasteleyn, C.M. Fortuin: J. Phys. Soc. Jpn. 26 (Suppl.), 11 (1969)
7.6   A.N. Burkitt, D.W. Heermann: Europhys. Lett. **10**, 207 (1988)
7.7   A.N. Burkitt, D.W. Heermann: Comput. Phys. Commun. **54**, 201 (1989)
7.8   G.C. Fox, S.W. Otto: Phys. Today 37 (5), 50 (1984)
7.9   G.C. Fox, M.A. Johnson, G.A. Lyzenga, S.W. Otto, J.K. Salmon, D.W. Walker: *Solving Problems on Concurrent Processors*, Vol.1 (Prentice-Hall, Englewood Cliffs, NJ 1988)
7.10   D.C. Rapaport: Comput. Phys. Rep. 9, 1 (1988)
7.11   W. Hillis: Sci. Am. 108 (June 1987)
7.12   S.F. Reddaway: DAP - a distributed array processor, in 1st Annual Symposium on Computer Architecture (IEEE/ACM), Florida, 1973 (IEEE, New York 1973)
7.13   A.J.G. Hey: Comput. Phys. Commun. **56**, 1 (1989)
7.14   A.J.G. Hey: In PARLE '89 - *Parallel Architectures and Languages Europe*, Vol.II, ed. by E. Odijk, M. Rem, J.C. Syre, Lecture Notes Comp. Sci., Vol.366 (Springer, Berlin, Heidelberg 1989)
7.15   D.J. Pritchard, C.R. Askew, D.B. Carpenter, I. Glendinning, A.J.G. Hey, D.A. Nicole: In PARLE - *Parallel Architectures and Languages Europe*, Vol.I, ed. by J.W. de Bakker, L. Nijman, P.C. Treleaven, Lecture Notes Comp. Sci., Vol.258 (Springer, Berlin, Heidelberg 1987) p.278
7.16   M. Creutz: *Quarks, Gluons and Lattices* (Cambridge Univ. Press, Cambridge 1983)
7.17   K.C. Bowler, R.D. Kenway, G.S. Pawley, D. Roweth: *Occam 2 Programming Language* (Prentice-Hall, Englewood Cliffs, NJ 1984)
7.18   E. Brooks III, G. Fox, M. Johnson, S. Otto, W. Athas, E. DeBenedictis, R. Faucette, C. Seitz, J. Stack: Phys. Rev. Lett. **52**, 2324 (1984)
7.19   G.S. Pawley, G.W. Thomas: Phys. Rev. Lett. **48**, 410 (1982)
7.20   G.S. Pawley, K. Bowler, R.D. Kenway, D.J. Wallace: Comput. Phys. Commun. 37, 251 (1985)
7.21   D.W. Heermann, K. Kremer, P. Nielaba: In preparation
7.22   K. Kremer, K. Binder: Comput. Phys. Rep. 7, 260 (1988)
7.23   K. Binder, D.W. Heermann: *Monte Carlo Simulation in Statistical Physics: An Introduction*, Springer Ser. Solid-State Sci., Vol.80 (Springer, Berlin, Heidelberg 1988)

7.24   P.H. Verdier, W.H. Stockmeyer: J. Chem. Phys. **36**, 227 (1962)
7.25   N. Madras, A.D. Sokal: J. Stat. Phys. **50**, 109 (1988)
7.26   F.T. Wall, F. Mandel: J. Chem. Phys. **63**, 4592 (1975)
7.27   J. Batoulis et al.: J. Chem. Phys., to be published (1990)
7.28   D.W. Heermann, P. Nielaba, M. Rovere: Comput. Phys. Commun., to be published (1990)

Chapter 8

8.1   G.C. Fox, S.W. Otto: Phys. Today **37**, 50 (May 1984)
8.2   G.C. Fox, M.A. Johnson, G.A. Lyzenga, S.W. Otto, J.K. Salmon, D.W. Walker: *Solving Problems on Concurrent Processors*, Vol.1 (Prentice-Hall, Englewood Cliffs, NJ 1988)
8.3   D. Rigby, R.-J. Roe: J. Chem. Phys. **87**, 7285 (1987)

Chapter 9

9.1   C.R. Askew, D.B. Carpenter, J.T. Chalker, A.J.G. Hey, D.A. Nicole, D.J. Pritchard: Comput. Phys. Commun. **42**, 21 (1986)
9.2   I.M. Baron, P. Cavill, D. May, P. Wilson: Electronics 109 (Nov.17, 1983)
9.3   *Transputer Reference Manual* (Inmos, Bristol 1986)
9.4   *OCCAM - A Compiler Writer's Guide* (Inmos, Bristol 1986)
9.5   C. Jesshope: Parallel Comput. **8**, 19 (1988)
9.6   Supercluster: Hardware Reference Manual, Parsytec, Aachen (1988)
9.7   Computing Surface: Hardware Reference Manual, Meiko, Bristol (1987)
9.8   *Occam Programming Language* (Printice-Hall, Englewood Cliffs, NJ 1984)
9.9   R. Steinmetz: *Occam 2* (Hüthig, Heidelberg 1987)
9.10   G. Jones: *Programming in OCCAM* (Prentice-Hall, Englewood Cliffs, NJ 1987)
9.11   K.C. Bowler, R.D. Kenway, G.S. Pawley, D. Roweth: *Occam 2 Programming Language* (Prentice-Hall, Englewood Cliffs, NJ 1984)
9.12   R.J. Glauber: J. Math. Phys. **4**, 294 (1963)
9.13   B. Lubachevsky: J. Comput. Phys. **75**, 103 (1988)
9.14   C.A.R. Hoare: Commun. ACM **21**, 666 (1978)
9.15   K. Binder (ed.): *Monte Carlo Methods in Statistical Physics*, Topics Curr. Phys., Vol.7, 2nd edn. (Springer, Berlin, Heidelberg 1986)
9.16   K. Binder, D.W. Heermann: *Monte Carlo Simulation in Statistical Physics: An Introduction*, Springer Ser. Solid-State Sci., Vol.80 (Springer, Berlin, Heidelberg 1988)
9.17   D.W. Heermann: *Computer Simulation Methods in Theoretical Physics*, 2nd edn. (Springer, Berlin, Heidelberg 1990)
9.18   M.H. Kalos: *Monte Carlo Methods* (Wiley, New York 1986)
9.19   W.G. Hoover: *Molecular Dynamics*, Lect. Notes Phys., Vol.258 (Springer, Berlin, Heidelberg 1986)
9.20   K. Kawasaki: *Phase Transitions and Critical Phenomena*, Vol.2 (Academic, New York 1972)
9.21   E.W. Dijkstra: Commun. ACM **18**, 453 (1975)
9.22   L. Onsager: Phys. Rev. **65**, 117 (1944)
9.23   H.J. Herrmann: J. Stat. Phys. **45**, 145 (1986)
9.24   M. Creutz: Phys. Rev. Lett. **50**, 1411 (1983)
9.25   M. Creutz: Phys. Rev. Lett. **63**, 9 (1989)
9.26   R.C. Desai, D.W. Heermann, K. Binder: J. Stat. Phys. **53**, 795 (1988)

9.27 S.F. Reddaway: DAP - a distributed array processor, in 1st Annual Symposium on Computer Architecture (IEEE/ACM), Florida, 1973 (IEEE, New York 1973)
9.28 R.W. Gostick: ICL Tech. J. 1, 116 (1979)
9.29 M. Creutz, L. Jacobs, C. Rebbi: Phys. Rev. Lett. 42, 1390 (1979); Phys. Rev. D 20, 1915 (1979)
9.30 R. Zorn, H.J. Herrmann, C. Rebbi: Comput. Phys. Commun. 23, 337 (1981)

Appendix

A.1 M. Creutz, L. Jacobs, C. Rebbi: Phys. Rev. Lett. 42, 1390 (1979); Phys. Rev. D 20, 1915 (1979)
A.2 R. Zorn, H.J. Herrmann, C. Rebbi: Comput. Phys. Commun. 23, 337 (1981)

B.1 A. Milchev, K. Binder, D.W. Heermann: Z. Phys. B 63, 521 (1986)
B.2 R.C. Tausworth: Math. Comput. 19, 201 (1965)
B.3 S. Kirkpatrick, E.P. Stoll: J. Comput. Phys. 40, 517 (1981)
B.4 W. Paul, D.W. Heermann, R. Desai: J. Comput. Phys. 82, 487 (1989)

Subject Index